Title Page

Rev. 09/06/2023

Feedback: alsnewideas@gmail.com

AF879440

Other science books by Alan Sewell:

https://www.amazon.com/Einsteins-Car-Relativity-Alan-Sewell-ebook/dp/B0CBKY7HW1/

Contents

Quantum Shockwave

Niels Bohr Albert Einstein Richard Feynman

"Those who are not shocked when they first come across quantum mechanics cannot possibly have understood it," said Niels Bohr, one of the few physicists who ever prevailed in argument with Albert Einstein.

"I like to believe the moon is there even when I'm not looking at it," protested Einstein. "I have thought a hundred times as much about the quantum problem as I have about general relativity theory. I cannot seriously believe in [quantum theory] because ...physics should represent a reality in time and space, free from spooky action at a distance."

"I think I can safely say that nobody understands quantum mechanics," concurred Richard Feynman, a father of nanotechnology and quantum computing.

This book presents the often shocking history, philosophy, controversies, and personalities of quantum mechanics. It explains in popular terms the great debates between Albert Einstein, who was skeptical of quantum theories, and QM's founders, like Niels Bohr. These unresolved controversies are debated as intensively by physicists today. My goal is to condense the meaningful information you might obtain from years of stressful reading of QM books into a day's interesting read. I've referenced QM videos (current through 2023) and recent popular QM books, so you can dig deeper into QM studies if you want, without having to wade through hundreds of books and Youtube videos to find the best ones that are scientifically vetted and interesting for a lay audience.

The book originated while I blogged with quantum physicists about quantum computing, billed as the emerging technology for solving complex problems beyond the computational power of existing machines. The hype surrounding quantum computing has since diminished, now that we know its engineering requirements are too formidable for today's technologies. Perhaps it will follow the development path of television, whose electromechanical incarnation transmitted by telegraph wires was patented in 1884 but had to wait for the development of over-the-air electromagnetic propagation six decades later to become a commercially practical device.

In the meantime, the alluring mysteries of quantum mechanics remain unfathomed. More than a century after its conception, some aspects of QM still haven't been well considered or well presented to the public. We're about due for another revolutionary advance in our understanding of these theories that describe the foundations of the Universe. We're hearing about quantum computing, quantum memory, quantum multiple realities, and other emergent topics of quantum theory. The public would do well to glean a heads-up understanding of the fundamentals of quantum mechanics now, before the next dam bursts.

Because the internal engine of quantum mechanics remains obscure, QM has become to New Age spiritualist writers what snake oil is to quack doctors. Yet there seems to be a kernel of legitimate science embedded in the quackery. Sir Roger Penrose, a pioneer of new fields of research in math, physics, biology, and cosmology, including collaboration with the late Stephen Hawking, says there is room enough in science to encompass discussion of the fantastic speculative aspects of quantum theory:

But what about fantasy? Surely this is the very opposite of what we should be striving for in science. Yet we shall be seeing that there are some key aspects to the nature of our actual universe that are so exceptionally odd...that if we do not indulge in what may appear to be outrageous flights of fantasy, we shall have no chance at all of coming to terms with what may well be an extraordinary fantastical-seeming underlying truth.

I'm interested in the scientific and metaphysical viewpoints and will therefore explain both. I'm a reviewer of books on physics, brain science, evolution, and artificial intelligence. My Amazon review of ***Quantum Enigma*** by the late Dr. Bruce Rosenblum generated a wide-ranging discussion of QM topics:

https://www.amazon.com/Quantum-Enigma-Physics-Encounters-Consciousness/dp/0199753814/

- *Wow! That review was awesome! I really appreciate it. I also intend to print it to have on hand as I read the book.*
- *I have already trudged through probably 20 other quantum physics books, and I kinda-sorta learned more from this review than many of my books.*
- *A great review! I've read perhaps two dozen books on am and still can't wait to read this one.... Loved your review.*
- *Really great summary- really helps to see the woods before possibly getting lost in the trees.*
- *Loved this review, the comments, and the book.*
- *What a stunning review! Thanks Alan Sewell. This is how all reviews should be.*

The kudos on that review inspired me to write this book, incorporating my prior studies of quantum mechanics going back to advanced studies in high school and continuing with university physics, and through current discussions with physicists who originated the concept of quantum computers. However, I wanted to do more than just rehash the scientific ideas commonly found in other books. I wanted to imbed the history, philosophy, and spirit of QM studies with the science in an integrated whole. Dr. Rosenblum reminds us that:

Presenting this material to nonscientists is the intellectual equivalent of allowing children to play with loaded guns.

With that caveat in mind, I will seek to explain the various interpretations of quantum mechanics clearly enough for you to form a reasoned opinion of their validity, while covering the interesting metaphysical questions wrapped around the interpretations.

The Metaphysical Stakes

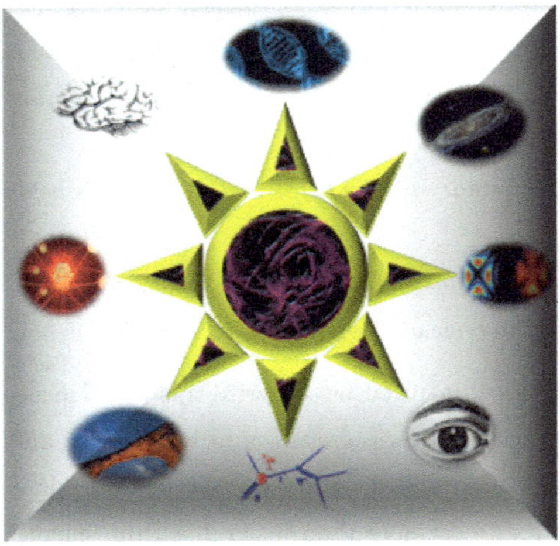

From its inception, quantum mechanics has been wrapped around seemingly remote subjects like cosmology, biology, and brain science --- especially the study of human consciousness. These controversies have escalated in recent years, as QM has become a foot soldier in the metaphysical war over whether our minds are entities created separately from the material universe, or whether they are merely a product of it. Reputations, careers, and untold "ego capital" is invested in extrapolating these opinions far beyond what science can prove.

Quantum mechanics became the wagon to haul these metaphysical controversies around in, because QM operates on the creation of reality from information, and information is what we process with our conscious minds. Nobel Prize winner Max Planck, whose discovery of "Planck's constant" became the fundamental measure of quantum processes wrote:

As a man who has devoted his whole life to the most clear-headed science, to the study of matter, I can tell you .... I regard consciousness as fundamental. I regard matter as derivative from consciousness.... Everything that we talk about, everything that we regard as existing, postulates consciousness. We have now discovered that there is no such thing as matter; it is all just different rates of vibration designed by an unseen intelligence.

Erwin Schrodinger. QM's preeminent founder, took up the question of consciousness in his book ***What is Life***:

https://www.amazon.com/What-Canto-Classics-Erwin-Schrodinger-ebook/dp/B00E3URD7Q/

Are we prepared to believe that this very special turn in the development of the higher animals, a turn that might after all have failed to appear, was a necessary condition for the world to flash up to ***itself in the light of consciousness?*** *Would it otherwise have remained a play before empty benches, not existing for anybody, thus quite properly speaking not existing? This would seem to me the bankruptcy of a world picture. The urge to find a way out of this impasse ought not to be damped by the fear of incurring the wise rationalists' mockery.*

Schrodinger, Erwin. What is Life? (Canto Classics) (p. 94). Cambridge University Press. Kindle Edition.

A boisterous coterie of quantum physicists believes quantum mechanics is the link between our consciousness in this world and the next. QM physicists and mathematician Sir Roger Penrose speculates on the biological retention of information in the brain, and the possibility of it being passed to a quantum soul:

https://www.express.co.uk/news/science/728897/LIFE-AFTER-DEATH-consciousness-continue-SOUL

LIFE AFTER DEATH: Shock claim of evidence showing consciousness may continue as a SOUL

...According to some well-respected scientists, quantum mechanics allows consciousness to live on following the body's eventual demise.

.... British physicist Sir Roger Penrose agrees and believes he and his team have found evidence that protein-based microtubules – a structural component of human cells – carry quantum information – information stored at a sub-atomic level.

Sir Roger states if a person temporarily dies, this quantum information is released from the microtubules and into the universe.

However, if they are resuscitated the quantum information is channeled back into the microtubules and that is what sparks a near death experience.

Sir Roger added: "If they're not revived, and the patient dies, it's possible that this quantum information can exist outside the body, perhaps indefinitely, as a soul."

Reposing the residue of our conscious minds in Eternity is not enough to satisfy some quantum theorists. Recent interpretations of quantum theory have centered around the "Many Worlds Theory" whereby a multiplicity of universes are theorized to be spawned by our conscious decisions:

https://www.newscientist.com/article/mg22329880-400-multiverse-me-should-i-care-about-my-other-selves/#

New Scientist Magazine

Every decision you make may spawn parallel universes where people are suffering because of your choice. Welcome to the quantum moral maze

I'm rich. I'm a movie star. I'm king of the world. I'm also poor. I'm homeless. Lots of me are dead.

This worries me. If many worlds is correct – and many physicists think it is – my actions shape the course not just of my life, but of the lives of my duplicates in other worlds.

https://www.nbcnews.com/mach/science/weirdest-idea-quantum-physics-catching-there-may-be-endless-worlds-ncna1068706

The weirdest idea in quantum physics is catching on: There may be endless worlds with countless versions of you.

Quantum mechanics incites these metaphysical vs. rationalist controversies in ways no other field of science does. It's a question of whether consciousness governs the Universe, or whether our perceived consciousness is a product of the Universe that only seems special because we experience it. I've chosen to explore quantum mechanics from the metaphysical intersection between consciousness and the theory of conservation of information. Does information ever disappear from the Universe, or is it a conserved quantity that once created exists forever?

This matters in quantum mechanics because some quantum theories postulate that objects do not become real (manifest a physical presence that affects other objects in the Universe) until information is obtained by observing them. According to these interpretations, we create objects when we observe them in the probability field of their pre-material existence. When we observe the probability field, we create the object and place information about it in our minds. If the information in our minds disappears when our brains die, does the object also disappear and return to its ghostly pre-existence? If the object maintains a permanent existence, does the information we gleaned by observing it remain permanent?

Since our memories are the collection of all we have observed and contemplated, must they remain permanent when our brain dies? If information obtained by observation never goes away, then all the information we've obtained in life never goes away. It must be reposed somewhere, perhaps in an immortal quantum soul.

Can we devise a scientific experiment that conclusively proves the Penrose theory that quantum information created by a person's observations during life can be accessed after that person's physical body dies? And if we did, what would be the consequences?

"We cannot make the mystery go away"

Quantum mechanics had its genesis in an experiment originated by Thomas Young, a British researcher of many fields of science and humanities, over two hundred years ago. The experiment requires only a light source, a board with two slits, and a screen on the other side to catch the light that passes through the slits:

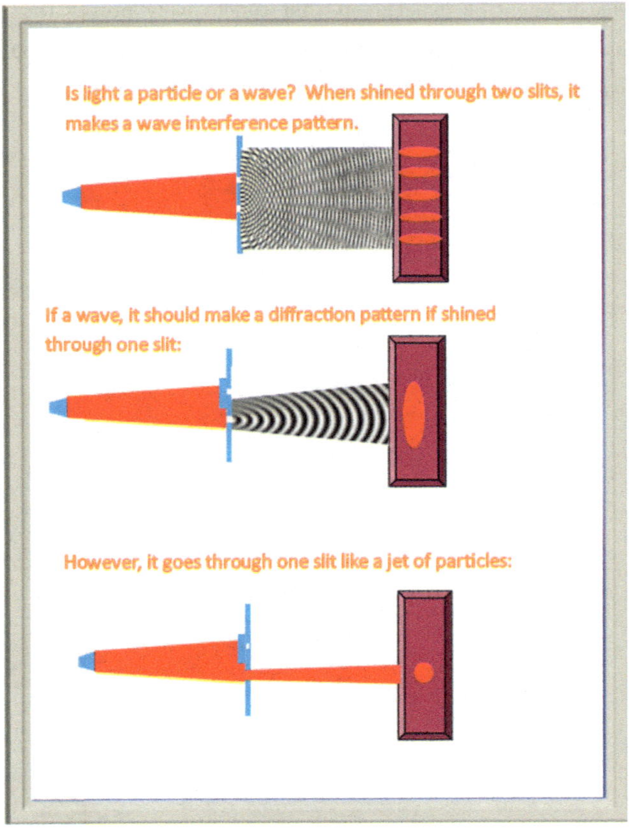

When Young reported the first "double slit" experiment in 1801, the scientist who later became the Lord High Chancellor of Great Britain decried it as "destitute of every species of merit" and "the unmanly and unfruitful pleasure of a boyish and prurient imagination."

What did Mr. Young do to provoke such outrage from a country noted for its culture of ***understatement?*** He showed that light manifested a dual particle / wave nature, but not in the classical way physics then understood waves to be. Young's experiment turned out to be the taproot of quantum mechanics. The experiment remains as inexplicable now as then.

[The two-slit experiment] contains the only mystery. ***We cannot make the mystery go away*** *by "explaining" how it works ... In telling you how it works we will have told you about the basic peculiarities of all quantum mechanics.... [the] one experiment which has been designed to contain all of the mystery of quantum mechanics, to put you up against the paradoxes and mysteries and peculiarities of nature."*

— Nobel Prize for Physics winner Richard Feynman

Young showed that when light goes through two slits, it looks like the familiar pattern waves make when an object is splashed in water or makes a noise in air. Water and soundwaves are propagation waves of stationary water and air molecules transmitting energy by bumping into each other. Waves of this nature interfere with each other when emitted from two sources. At some points the waves manifest constructive interference where their crests combine to make larger crests, while troughs combine to make deeper troughs. At other points, there is destructive interference where crests and troughs cancel. Noise cancellation devices work by emitting "anti-noise" signals out of phase with environmental noise, so that sound waves cancel by destructive interference. Light makes the same constructive and destructive interference patterns when it goes through two slits, thus bolstering the theory that it travels as a wave.

Then Young blocked one of the slits, expecting the wave behavior to continue, as shown in the middle of the picture. However, the wave behavior vanished. Light shot through the single opening like a jet of water moving through air. Light behaves like a wave when passing through two slits and like a jet of particles when passing through one.

Young's idea that light behaves as a wave was called "prurient" because Isaac Newton's scientific legacy was influential, and Newton had theorized that light travelled as particles. But perhaps light has a dual nature. Maybe it travels as particles through space, generating electromagnetic fields as they move. Perhaps when light goes through two slits, the electromagnetic fields interfere with each other like waves of water, but when light goes through only one slit, the fields don't interfere, and the photons shoot across space as particles.

Light was eventually shown to move through space in precisely that way, as photons generating oscillating waves of electrical and magnetic fields as they move. This caused light to behave like a "classical" wave of water or sound that interferes with itself when it goes through slits. This type of interference is known as diffraction. However, the slits must be microscopic to make the tiny electromagnetic waves bend around objects and cause diffraction interference. This effect is not visible with large objects. If you place a physical barrier between yourself and the sun, you don't see light bending into the shadow. However, the noise from a passing bird that chirps behind the barrier isn't blocked because noise is a kinetic wave through a medium of air that bends around surfaces. The interference of light that Young saw through the double slits was a different phenomenon. To assume it was caused by diffraction of waves bending around the edges of the slits would be as fallacious as assuming that because waves of water are caused by wind, that tides and tsunamis originate in the same way.

In 1983, it became possible to fire photons through the slits one at a time. The one-off photons also generated interference. How could a photon interfere with itself? That could only happen if photons were travelling as spread-out "waves" much larger than the electromagnetic waves we already knew about. But if each photon is travelling as a classical wave, we'd expect most of it to impact the opaque barrier around the slits, while a smaller part of it would pass through to impact the measurement screen to make an interference pattern from little bits and pieces of each photon.

However, when photons are fired through the double slits one at a time, it is an all-or-nothing event. Photons are always detected as whole units that land at one point. Electronic devices and our eyes see them that way. Either the entire photon gets through the slits and lands in one whole piece on the other side, or none of it does. Most photons are stopped by the barrier. Those that get through the slits build up interference patterns one-by-one on the measurement screen, like they do when trillions go through the slit simultaneously in a light beam. Each photon particle must therefore be interfering with itself.

A photon interfering with itself is inexplicable by classic wave mechanics. However, photons are massless particles that travel at the constant speed of light without experiencing the passing of time. Could that be what allows a photon to interfere with itself? Then it was shown that the same thing happens with electrons, atoms, and even molecules made up of many atoms. These are particles with mass that travel at less than the speed of light and therefore experience the passing of time, so one particle can't be interfering with its past or future incarnations. Like photons, these particles of mass create interference patterns when going through two slits, while creating jets of particles going through one. It seems that all objects do this, up to some arbitrary, and still to be determined, size.

It must be that particles move across space as a "cloud" spread out over a wide area. When a particle-cloud encounters a barrier with two slits, it either materializes as an impact point on the barrier, or it passes through both slits as two clouds that spread out beyond the barrier and interfere with each other, thereby warping each particle's impact point into a pattern of interference bands that manifest themselves after many particles are fired through the slits one by one:

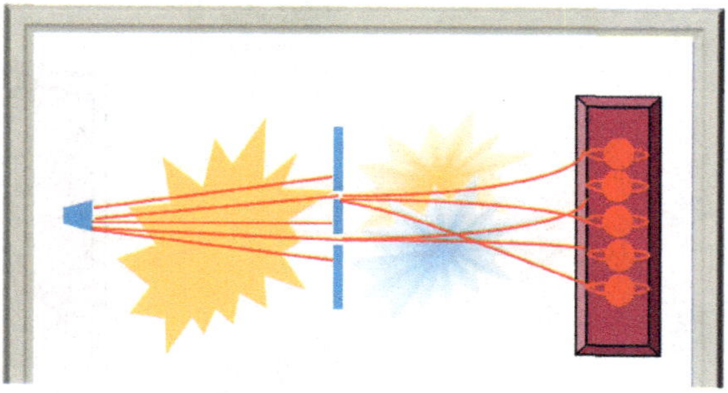

When a particle-cloud encounters a barrier with only one slit open, the particle either materializes as an impact point on the barrier, or it goes through the slit in a straight line without interference and materializes as an impact point on the measurement screen. When many particles are fired through the single slit one-by-one, they land close together, making a "clump pattern:"

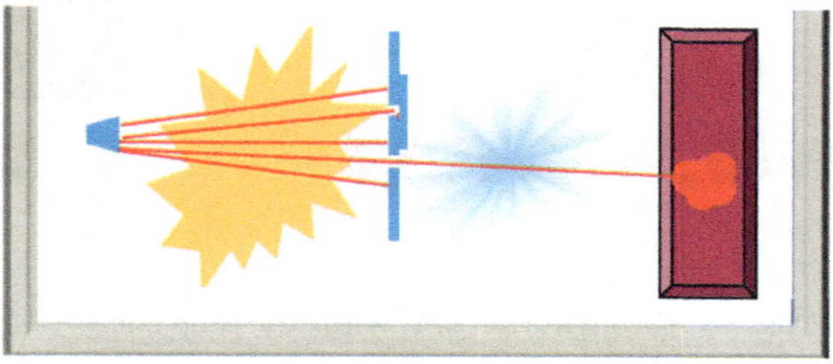

Now comes the heart of the mystery: We don't need to block one of the slits to eliminate the interference pattern and make the particles land in a clump. We can leave both slits open and eliminate the interference pattern just by identifying which of the slits each particle passes through. We could do this by placing a detector next to each slit, but it turns out we only need one detector placed next to one of the slits.

If the detector is turned off, as shown in the upper half of the picture below, the particles make an interference pattern. As soon as the detector is turned on, as shown in the bottom half, the interference pattern changes to a clump pattern:

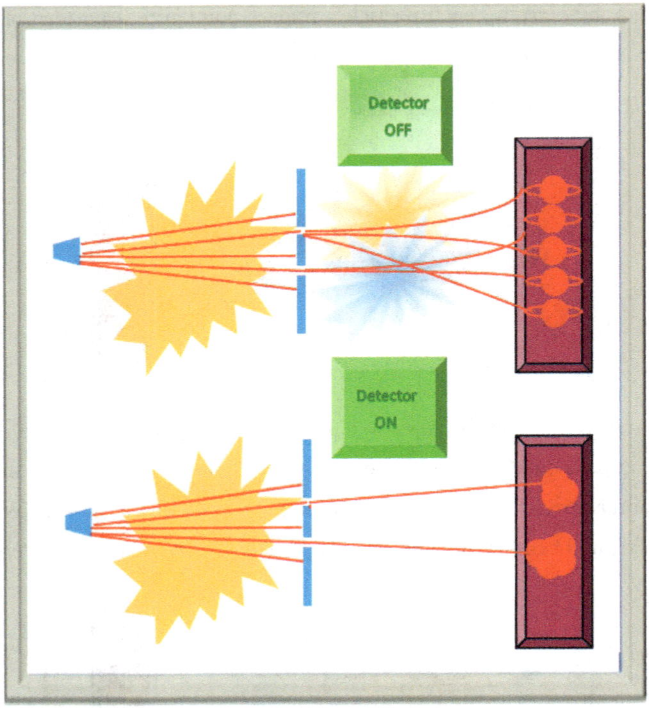

If the electron passes through the slit with the detector, its electric charge imparts information about its location to the detector that reveals the electron's position and causes it to materialize into reality as a straight-line particle track that doesn't interfere with itself. An interaction has taken place whereby the electron's electric field has influenced the electrons in the detector, causing it to send a signal to a memory device that records the electron's passing. We might theorize that the detector has influenced the electron in some way that causes the electron to materialize.

Now comes the mystery If the electron passes through the slit that ***doesn't*** have the detector, then no interaction takes place. Yet the electron also materializes into a particle that doesn't interfere with itself. This is because we can ***infer*** that if the electron got to the measurement screen without activating the detector, it had to pass through the slit without the detector.

Thus, it seems that it isn't interaction, but rather ***information***, including ***inferred information,*** of which path a particle takes through the slits that converts it from a cloud into a particle with definite properties even if it does not interact with a detector.

This ability to manipulate a particle by controlling the information we know about it implies that we may have the power to control some aspects of the Universe with our minds by choosing how we observe it. Some interpretations of quantum mechanics postulate a quantum chain reaction from the smallest particles to the largest, all the way up to the Universe itself.

This notion of changing the Universe by gleaning information about subatomic particles sounds preposterous until we think about how we can initiate stupendous chain reactions in hydrogen bombs. A few quarts of hydrogen are enough to destroy a large metropolitan area after a fusion chain reaction in one hydrogen atom is ignited. We could, with modest effort, chain enough hydrogen atoms together to blow the surface of the Earth off into space. If we can manipulate something tiny by controlling how much information we know about it, there may be no upward limit to the chain reaction we can manipulate. Because our minds store information, we come to the question of whether they have power to shape some aspects of the Universe.

Let's theorize that our brains might operate on two levels:

Level I. Our brains are electromechanical "machines" that store information like a computer memory. Our minds are merely the recall of information stored in the brain as each organism's current experiences layered on top of instincts honed by billions of years of blind evolution. If this is the highest level our brains are capable of, then consciousness is our mind's illusion to make us think we are making decisions based on free will when we aren't.

Level II. Our consciousness is a product of the mind, separate from the Universe of matter, energy, space, and time. Perhaps our minds are so separate from the material Universe that they cannot be explained by material processes such as the electrical impulses in our brains that affect its atoms and molecules.

The ten or so named interpretations of quantum mechanics fall into several fundamental categories, mostly based on how much, or how little, our mind's consciousness is theorized to be involved in storing information that triggers quantum events. The interpretations most widely debated by quantum physicists are:

- "Copenhagen" (in honor of Niels Bohr's physics lab). Since we can explain the results of quantum mechanics by mathematics, there is no need to care whether our minds play any role in creating the results. *The observer has, rather, only the function of registering decisions, i.e., processes in space and time, and it does not matter whether the observer is an apparatus or a human being; but the registration, i.e., the transition from the "possible" to the "actual," is absolutely necessary here and cannot be omitted from the interpretation of quantum theory. – Werner Heisenberg*

- Extreme Copenhagen or Consciousness Causes Collapse Hypothesis (CCCH). Our minds create the Universe, past, present, and future. *"We are participators in bringing into being not only the near and here but the far away and long ago. We are in this sense, participators in bringing about something of the universe in the distant past and if we have one explanation for what's happening in the distant past why should we need more? – John Wheeler*
- Many Worlds. Multiple Universes are spun off every time a decision is made consciously by a human being, or unconsciously by an inanimate physical process, to maintain the existence of all possible quantum states.
- Hidden variables. The idea that objects do not suddenly spring into existence when we observe them. They are there all the time in real form, driven through space by "hidden variables" that we don't know about yet. Until we find these "hidden variables" the only way we can see locate particles is to cast a wide net to catch their "cloud."

The double slit experiment seems to show that the acquisition of information controls the properties of small particles like photons and electrons, and that information can be inferred without being directly observed. When information is gained by being inferred, it is known as an ***interaction-free observation*** because the act of observation does not affect the object being observed by bombarding it with photons, electrons, or other particles; or by changing its velocity by accelerating it in a gravitational or electromagnetic field. We can thus transform waves into particles without interacting with them. This will be the way we test for whether the information stored in our brains remains a permanent part of the Universe after we die, when the physical matter of the brain is destroyed.

We do not know which interpretation of quantum mechanics is correct. We do know our brains are stores of information gleaned by observation and inference, regardless of whether that information is manipulated by instinct (level I) or conscious deliberation (level II). That should be enough to enable us to conduct our QM experiment to confirm or debunk the concept of the immortal soul.

Schrodinger's Wave

Erwin Schrodinger

Quantum mechanics is named for the mathematics that predicts the absorption and release of energy from atoms in discrete "quanta" packets transmitted when electrons change their positions and emit photons that carry energy to other atoms. Electrons make "quantum leaps" downward in their orbits when they lose energy by transmitting photons, and "quantum leaps" upward when their energy increases after they receive photons emitted by other atoms. They make all-at-once movements between defined positions in atoms, without stopping anywhere in between. That part of the theory is well understood.

The controversial aspect of quantum theory is the nature of the "wave" whose interference patterns we detect in the double slit experiment. We can make our journey into QM easier by clearing up some confusing ideas at the outset. The taproot source of confusion involves the "quantum wave function." Several types of waves are discussed in physics.

The most common wave we experience is the propagation of kinetic energy through a medium, such as energy transmitted through sound and water. These waves transmit energy from a source to an object that absorbs or reflects them. In air and water, propagation waves move by the transmission of energy via air and water molecules bumping into each other while remaining relatively stationary due to being tightly packed together. The particles remain where they were after transmitting their kinetic energy to neighboring particles.

Light, X-rays, radio, and other energies on the electromagnetic spectrum are transmitted by photons. As photons move through space at lightspeed, they generate electromagnetic forces on other particles that cause them to bob up and down. This is a different process than the waves we see in air or water that transmit kinetic energy when stationary particles molecules bump into each other.

Photons of low energy produce radio signals. Photons of moderate energy produce visible light. We usually refer to radio and light as "waves" because they were discovered in the late 1800s at a time when James Clerk Maxwell had developed mathematical formulas that shifted the weight of opinion toward the view that light travels primarily as an electromagnetic wave through a medium, like sound or water. The theoretical medium was called "aether" (to distinguish it from the anesthetic ether) that supposedly filled space. Numerous experiments failed to detect "aether," so the view began to shift back toward Newton's view that light travelled through space as a jet of particles. When particles more energetic than light were discovered, we began to speak of them as x-***rays*** and gamma-***rays*** because we knew they travelled in straight lines as particles.

Then in 1925 Erwin Schrodinger devised the Schrodinger Equation to describe quantum mechanics, as Newton invented the Laws of Motion to describe classical motion, and Albert Einstein and others invented math to describe Relativity. Schrodinger's equation became known as "the wave function" even though it has nothing to do with physical waves. Perhaps "wave" seemed appropriate because Schrodinger's mathematical formula, when plotted as a graph, resembles a series of sine waves.

Schrodinger's equation postulates that everything in the Universe, including the Universe itself, can be described by this mathematical "wave function" that plots peaks and valleys on a graph indicating the probability of finding an object having properties such as location in space, energy content, velocity, spin, and so on. The Schrodinger equation predicts how these probabilities evolve over time, and how likely the object is to have them when it is observed.

Here is a Youtube video showing graphic depictions of the mathematical "wave function" of a particle moving in a confined space, between the walls of a box:

https://youtu.be/g4wuSgwLT9I?t=83

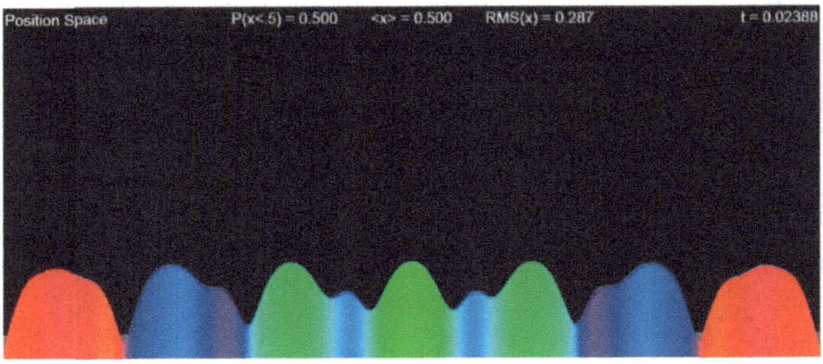

Until it is observed, the particle doesn't exist at any specific location inside the box. The height of the mathematical "wave" at each point is the probability that the object will be discovered at that point in space if someone looks for it inside the box. The particle will more likely be found at the peaks of the "wave" than its troughs. Other aspects of the particle, such as its energy content, are also mathematical probabilities. The particle doesn't become definite until it is observed and measured by something that already exists.

The particle's existence is so undefined that the phenomenon of "quantum tunneling" takes place whereby a particle's probability wave can pass through barriers that the particle itself cannot penetrate. In the picture below, particle detectors are placed in front of and behind a barrier. When the detector materializes a particle before it encounters the barrier, the particle impacts the barrier and cannot go through it. When the detector is placed behind the barrier, some of the probability wave can pass through it, making it possible that the particle will materialize on the other side of it, as if the barrier wasn't there:

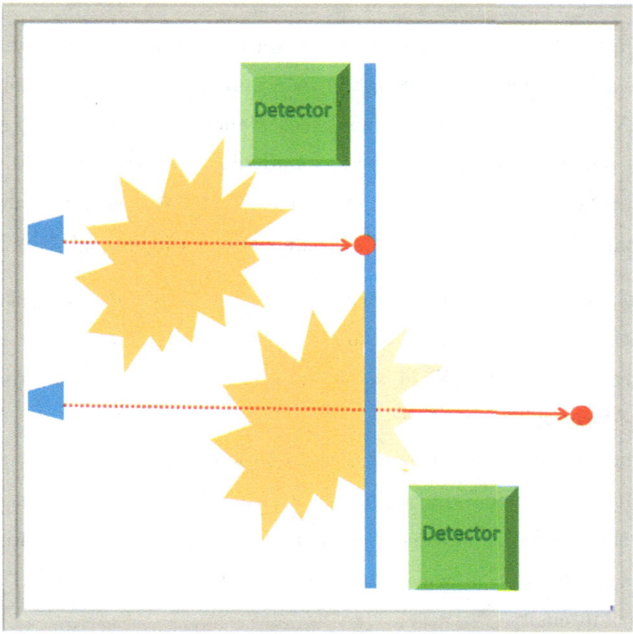

In fact, the particle isn't anywhere before it is materialized by detection. It is a wave of abstract probabilities of where the particle might be discovered if we go looking for it. The particle will probably be discovered close to its expected line of trajectory, but its probability wave attenuates into low-order possibilities that the particle can be found on Mars, Andromeda, or anywhere in the Universe. If the particle has become "entangled" by interacting with another particle in the past, then both will materialize simultaneously into existence when one of them is observed, even if they happen to be on opposites sides of the Universe.

We can define a probability wave for practically any phenomenon. Schrodinger's equation doesn't apply to predicting the characteristics of human beings, but we can use its principle to plot a probability wave of humans with certain characteristics.

For example, we could plot "the probability wave of MLB baseball fans." The baseball wave will manifest a cyclical nature as it travels through time. It will be dormant every year from November through March. Around April it will start stirring to life as baseball fans appear at MLB games. The wave will show cyclical peaks on Saturday and Sunday afternoons and on weeknights, when most games are played. It will show troughs during weekdays while most fans are working their day jobs. It will show areas of concentration at places such as MLB stadiums on game days, and areas of smaller concentration around TV sets as fans watch the games at home.

We can compute, with a fair degree of certainty, how much of the wave will be hovering around MLB games at any given time during the year, and how much will be hovering around sports bars and at-home TV sets. We can forecast how much of the wave will be found at any given place at any given time, but not the specifics of where any individual fan would be, unless we already knew something about the fan. Think of the sum of all baseball fans as a probability wave, and the individual fans as particles in the wave. Thus, we can speak of the baseball fans in a couple of ways:

1. As a probability wave (quantum wave function) that moves from place to place depending on what season it is and on what day and time it is. The wave is more likely to be concentrated at Wrigley Field at 2:00 PM on a Saturday in July than at 3:00 AM on a Sunday Morning in December.
2. As a propagation wave of people getting to the stadium by car or bus, or by getting to a game on TV in their living rooms by walking from the refrigerator to the sofa. Even though the people are moving as particles, it is intuitive, though misleading, to think of them as a propagation wave phenomenon, like sound or water, that has peaks and troughs of intensity.

To avoid confusion between these two types of waves, I will refer to quantum probability waves, defined as mathematical formulas by Erwin Schrodinger, as ***"Schrodinger's waves."***

Quantum Observation and Inference

The act of observing or measuring a quantum particle's Schrodinger's wave seems to create a reality out of something that was a mathematical abstraction an instant before. As Pascual Jordan, one of the founders of quantum theory, wrote:

"Observations not only disturb what is to be measured, they produce it."

Until then, the object-in-waiting is a probability wave consisting of a "superposition" of all possible properties and locations. Let's consider this quantum superposition principle as it might be applied to baseball fans.

Let's assume every baseball fan on the planet is a superposition of just one fan. When we scan the world to find that one baseball fan, all the millions of superpositions will go away and only one real one will be left. Let's imagine we have a satellite in orbit programmed to detect that one baseball fan. The satellite has a wide-angle camera that can photograph several hundred square miles of the Earth's surface each second as it sweeps around its orbit. At some point during this observation, the Schrodinger's wave of all possible baseball fans will "collapse" into a single living, breathing individual. The other millions of superposition fans will disappear as if they never existed.

Quantum mechanics is often mis-taught as the Schrodinger's wave being collapsed by the intrusive action of the observer --- for example, the observer disturbing the space around the object by shooting electrons or photons through it, to discover the object by recording electrons or photons it reflects. This is an incorrect theory because the object does not yet exist, so it cannot be affected by particles we are sending into its space. Moreover, the observer, i.e., the satellite camera in this example, is passive. It merely observes the space where a baseball fan might be located, and the act of surveilling that space causes one to materialize.

As we've seen, the observation can also be inferred by interaction-free observations. If it so happened that we were looking for an MLB fan in Chicago who had season tickets to attend every home game, and if both teams are playing at home simultaneously that day, we could discover that fan by inference. If we looked for the fan at the Cubs' Wrigley Field and he/she wasn't there, he/she would materialize at the White Sox' park without us having to look there. The fan can only be in two places on that game day, so we only need to look for him / her in one of them, just like we only need a detector at one of the slits to know which one the electron or photon went through. Once we've identified the space the fan is in, either by direct observation or inference, a single fan with myriads of specific characteristics materializes there, and all the other superposition fans go away.

The ability to harness the superposition principle of quantum mechanics to obtain information about an object without directly observing it has recently been popularized in the *Elitzur–Vaidman bomb tester experiment* whereby an unexploded bomb might be identified without looking at it any direct way that would detonate it:

https://en.wikipedia.org/wiki/Elitzur%E2%80%93Vaidman_bomb_tester

I prefer a more peaceable analogy, so I've applied the principle to observing the mythical creature "Bigfoot," by assuming Bigfoot is real, rather than merely a dubious legend derived from Northwest Canadian Indian lore in the 1920s. Let's see if we can harness quantum mechanics to find him.

Bigfoot is said to be so wary of humans that the only time he's ever been "seen" is when human beings dress up in monkey suits to emulate him. We want to see the real Bigfoot, without alerting him to our presence. In normal life we could do this by placing hidden cameras, motion detectors, and so on, around Bigfoot's presumed stomping grounds. But maybe Bigfoot is a super-intelligent alien creature long practiced in detecting the electric fields from manmade devices. We want to detect him discretely without leaving any telltale manmade devices or detectable electromagnetic fields in his path. How can we do it? Maybe we could put a telescope on a mountain or on a space satellite and see him that way, but let's assume that Bigfoot knows about those and can camouflage himself. We want to be certain we can detect him, and he can't detect us. This is not possible with classical physics, but as Avshalom Elitzur and Lev Vaidman demonstrated in their bomb tester experiment, quantum mechanics gives us a way to infer the information remotely without being detected. Let's devise the following apparatus:

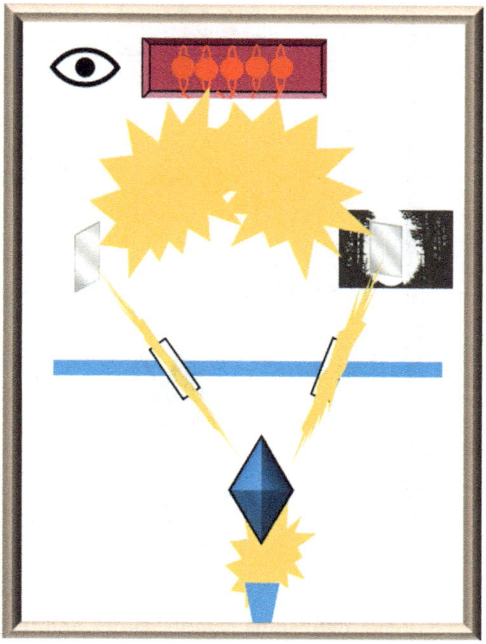

We've placed a source of photons in front of a device known as a beam-splitter. The beam splitter provides a more directed path through the double slits, so we don't lose any photons or electrons impacting on the metal bar that separates them. The beam splitter has a random 50/ 50 chance of sending a photon either way, so we don't know which path the particle has followed when it hits the measurement screen. We've placed two mirrors behind the slits to redirect the two possible paths to converge on a measurement screen a ways off where Bigfoot can't see it. The mirror on the right is behind the path we think Bigfoot walks. Of course, Bigfoot can see the mirror, if he looks that way, but perhaps he's a vein creature that likes to preen in front of it and won't run away from it.

The light source emits one photon at a time. If Bigfoot isn't walking the path, each photon takes the superposition of both paths and interferes with itself, as shown above, thereby generating the interference pattern on the measurement screen. Then comes Bigfoot sashaying down his trail, blocking the photon's path on the right. Because the beam splitter has a 50% chance of sending the photon either way, one of two things can happen. If the photon goes to the right it impacts Bigfoot:

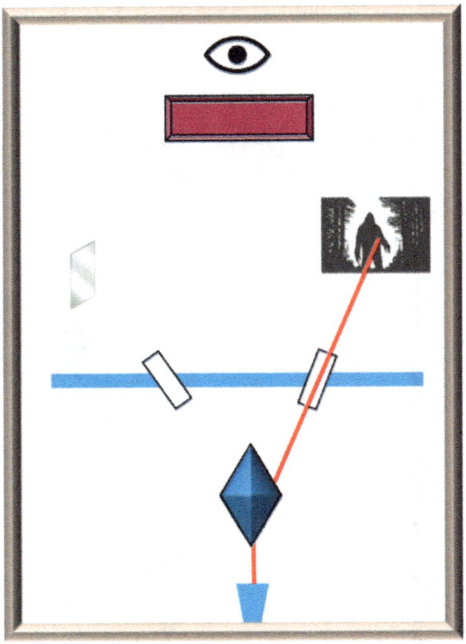

The photon's path becomes defined back to its source and the measurement screen goes dark because the photon can't get there. We know Bigfoot is blocking it, but this is not a useful outcome for us, since Bigfoot presumably detected the photon and scampers away, just like he would if we shined a flashlight in his face. The next alternative, whereby the photon takes the left path is the one that provides useful information:

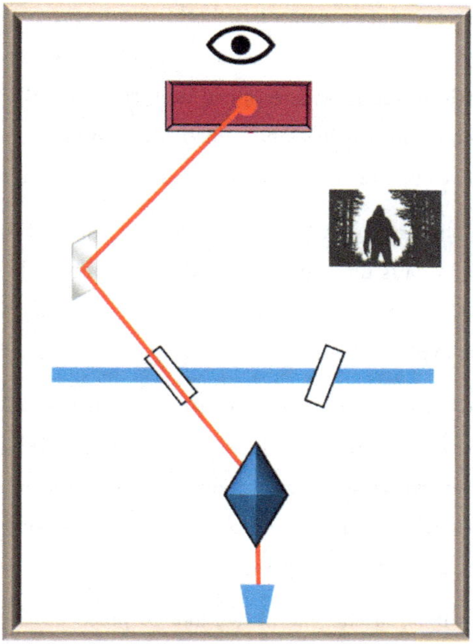

In this case, the photon only has one path to the measurement screen because Bigfoot is blocking the one on the right, so it lands on the screen in a non-interference position. We know Bigfoot is there because the photon got to the measurement screen without interfering with itself. However, Bigfoot has no way of knowing that we detected him, since he didn't sense the photon. This is not a perfect method of detection, because there's that 50% chance we're going to scare him off if the photon takes the right path and impacts him directly. Still, there's a 50% chance of us knowing he's there without him knowing he's under observation. If we're lucky enough to get that 50% chance of detecting him, without him detecting us, then we have obtained some useful information. We know where he is at this moment, so maybe we can hop over there and take an up close and personal picture, or even engage him in a friendly palaver.

Avshalom Elitzur and Lev Vaidman developed this experiment using bombs instead of Bigfoots. Suppose an enemy has air-raided one of our cities. As often happens in war, the enemy wants to inflict maximum physical and psychological damage with minimum use of resources. He has dropped a thousand bombs on our city. Half are dummies with no explosives or fuses. The other half are on time delay fuses set to detonate hours, days, weeks, or months in the future. They are set to explode if disturbed by touch or X-rays that try to peer inside them.

Look at what the enemy has accomplished: He has doubled the number of bombs he can deliver. He only had enough fuses and explosives to make 500 real bombs, so he added another 500 dummy bombs that cost little to manufacture, since they require only an empty shell. Because we don't know which of the thousand are live and which are dummies, we must assume they are all dangerous. How do we remove or at least neutralize the bombs without detonating the live half?

We could evacuate the area and try to detonate all the bombs from a safe distance. This would result in 500 explosions, causing heavy destruction of property, but hopefully no lost lives. The time required to detonate the bombs would be doubled, because we don't know which half of the 1,000 are live until we try to explode them.

An alternative approach would be to pour concrete shells over all 1,000 bombs to dampen the explosions of the live ones when their time delays set them off. This would require an immense amount of concrete and labor, which we need to allocate to the active battle fronts, thus degrading our war effort. Some bombs will detonate before the concrete is poured, causing destruction and perhaps killing the concrete crews in the process.

This is exactly the dilemma the enemy wanted to place us in. If we detonate the 500 live bombs, much property be destroyed, and people will complain: "You could have covered them up!" To forestall a catastrophic decline in civilian morale, the governing authorities might opt to cover the 1,000 bombs with concrete, thereby drawing personnel and material away from the battle front.

However, quantum mechanics offers us what might be considered a more optimal way out of the dilemma. Perhaps we can identify the live bombs by indirect quantum observation, without exploding them, the way we detected Bigfoot by inference. We could set the Bigfoot apparatus up to examine the bombs by sending a photon through them. If the bomb is a dummy, the same thing happens as in the scenario when Bigfoot isn't there to block the mirror. We get an interference pattern, telling us the bomb can be safely removed.

If the bomb is armed, the same two possibilities happen as in the Bigfoot-is-there scenario. There is a 50% possibility that the photon will travel the path to the right, impact the bomb's fuse, and detonate it. Massive destruction will occur, the same as would have happened if we'd chosen to detonate the bombs on purpose. There's also a 50% probability that the photon will take the path to the left and identify the bomb as being live without detonating it. Thus, we will know to pour concrete shells only around those bombs without wasting it on the rest.

So here are the three scenarios combined --- two classical physics scenarios and one quantum scenario:

Classical Physics Scenario #1: Attempt to detonate all bombs. Result: 50% of bombs are found to be dummies, and the other 50% detonate, causing widespread destruction, though hopefully no loss of life.

Classical Physics Scenario #2: Pour concrete around all bombs. Result: 100% of bombs get a concrete shell built around them, at huge expense of concrete and labor, thereby drawing resources away from the battle fronts. Furthermore, we can't pour concrete over every bomb all at once. Some will explode anyway, possibly killing the crews working to get the concrete over them.

Quantum Mechanics Scenario: We test every bomb with the Elitzur and Vaidman device. Half the bombs are revealed as dummies due to interference patterns showing both paths taken in superposition. Of the remaining half, half of those (25% of total) are detonated when a photon takes the path to the right and activates the fuse. The remaining half of the half (25% of total) are identified as live bombs when the photons take the path to the left and land as a dot with no interference pattern. Those live bombs have concrete poured over them.

We have thus improved by quantum inference our options over what classical physics could give us. We have reduced the explosion of live bombs by half, therefore limiting the damage to 25% of bombs dropped instead of 50%. We have reduced our allocation of concrete and labor by 75%. Because we can pour concrete over 25% of the bombs faster than over all of them, we reduce the risk to our concrete pouring crews. Our civil population is also happier. Instead of blasting their homes and places of business by detonating the live 50% of bombs, we have detonated only half of those. The civil population says: "Them's the breaks. At least they tried to save us all and did save half." Our war effort is less impeded, our civilian morale remains high, and we prevail.

These Bombs-and-Bigfoot scenarios are contrived, but perhaps we are not far off from useful real-life applications of quantum inference. For example, detecting a disease by quantum inference instead of bombarding it with x-rays that might damage healthy tissue.

This power of quantum inferred observation not only transcends space, but also time. Let's say that Bigfoot has a home base on Mars. You transport a mirror to Mars and set it up on Bigfoot's path, while leaving all the rest of your apparatus on Earth. If Bigfoot walks in front of the mirror on Mars, you'll get the same outcome as if he walked in front of the one on Earth:

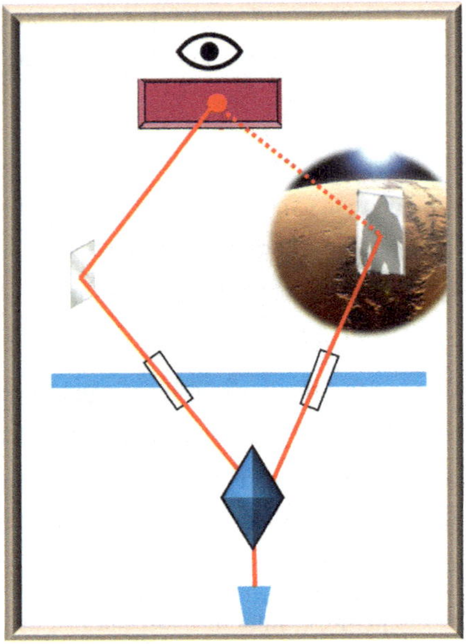

If Mars happens to be ten light-minutes away at that moment, then the photon will not impact Bigfoot (if he's there) or bypass him (if he's not there) for ten more minutes. Then it will take another ten minutes to make its way back to the measurement screen on Earth, assuming the Mars mirror is very carefully aligned to anticipate the measurement screen's future position as Earth moves through space.

If the mirrors are properly aligned, you will know immediately whether Bigfoot will be in front of the mirror on Mars in ten minutes, because your measurement screen is here on Earth, only a few light nanoseconds away from the source. The photon is travelling in superposition over both the long Mars-path and the short Earth-path. If the long path to the right is blocked by Bigfoot on Mars (or if it is unobstructed because Bigfoot is not there) when the photon arrives ten minutes later, then the photon will complete the short path on Earth, impacting the screen in nanoseconds. An event that will not happen until ten minutes in the future on Mars is known in the present moment of Earth. Is it any wonder Einstein revolted at the notion of this "spooky action at a distance?"

This effect has been demonstrated by Austrian physicist Anton Zeilinger whose team created a short-path / long-path experiment of photons in the Canary Islands. The long path was 144 kilometers and the short path a couple meters. At the end of the 144-kilometer-long path, a random number generator was used to electronically obstruct or not obstruct the path. The photon impacting the measurement screen on the short path predicted whether the long-path photon would be randomly obstructed at the end of its 144-kilometer flight path before the photon got there! This experiment is described in detail in the book ***Through Two Doors at Once*** by Anil Ananthaswamy.

https://www.amazon.com/Through-Two-Doors-Once-Experiment-ebook/dp/B079KTT95K/

The superposition principle seems to have transcended space and time. Now we'll see if it transcends death.

The Reality Chain

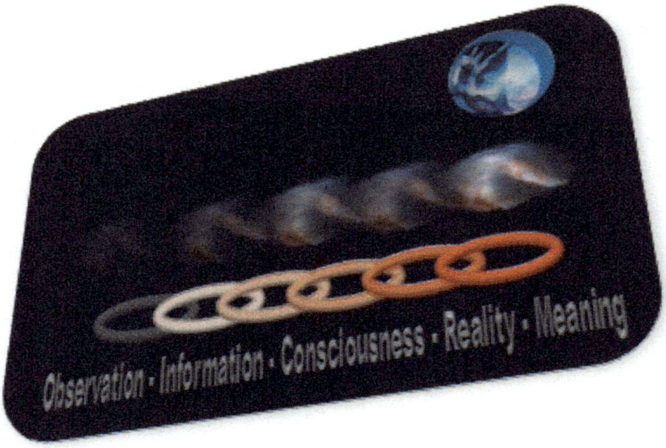

The concept of the "reality chain" originated with mathematician John von Neumann. His eponymous ***Von Neumann Chain*** postulates that a Schrodinger's wave superposition becomes a physical reality at some point between the time it encounters a materialized object larger than itsel and when it is observed by a human mind that understands what it is observing.

Von Neumann's interpretation of quantum mechanics was given credence based on his formidable reputation. He pioneered the designs of computers, atomic weapons, the "von Neumann probe" of self-replicating spaceships, and even some aspects of modern economics, which might have won him a Nobel Prize in that field had he not died in his early 50's. He originated the concept of the "technological singularity," a theoretical point at which we might artificially amplify our intelligence to the point where it increases exponentially toward infinity.

"It seems fair to say that if the influence of a scientist is interpreted broadly enough to include impact on fields beyond science proper, then John von Neumann was probably the most influential mathematician who ever lived," wrote a colleague.

Although von Neumann never claimed that human consciousness is the *only* point where a Schrodinger's Wave is collapsed into reality, the fact that he said it *might be one* of the points was enough to be expanded upon by other QM founders John Wheeler, Niels Bohr's son Aage Bohr Henry Stapp, and Eugene Wigner. They hardened Von Neumann's view by claiming tha Schrodinger's waves can *only* be collapsed into materialized objects by observations filtered through human consciousness. Their view is summarized as

There exist external observers which cannot be treated within quantum mechanics, namely human (and perhaps animal) minds, which perform measurements on the brain causing wave function collapse.

They theorized that inanimate measuring devices like cameras and Geiger counters cannot cause microscopic particles to materialize from Schrodinger's waves into realities because those devices are made from atoms that also exist as Schrodinger's waves. Their quantum mathematics showed, at least to their satisfaction, that if a particle in Schrodinger's wave superposition is observed by any material device --- a measurement screen, a photon / electron detector, camera, human eye and brain, or whatever --- then those devices are put into superpositions that do not cause Schrodinger's waves to collapse into realities. They believed Schrodinger's waves can only be converted into realities by nonmaterial consciousness that is aware of itself, aware of what it is looking at, and aware of the meaning of what it is looking at. This view seems so strange that we may wonder how it was taken seriously by physicists whose pragmatic minds created devices such as atomic bombs and computers.

This view was never held by most physicists and further declined in credibility as knowledge of quantum mechanics matured. John Stewart Bell was especially boisterous in opposing the notion that consciousness played any special role in creating reality. He drew a diagram of his conception of the reality chain, which I have edited for clarity, and then juxtaposed with the CCCH (Consciousness Causes Collapse Hypothesis) view that Schrodinger's waves can only be collapsed by sentient human minds:

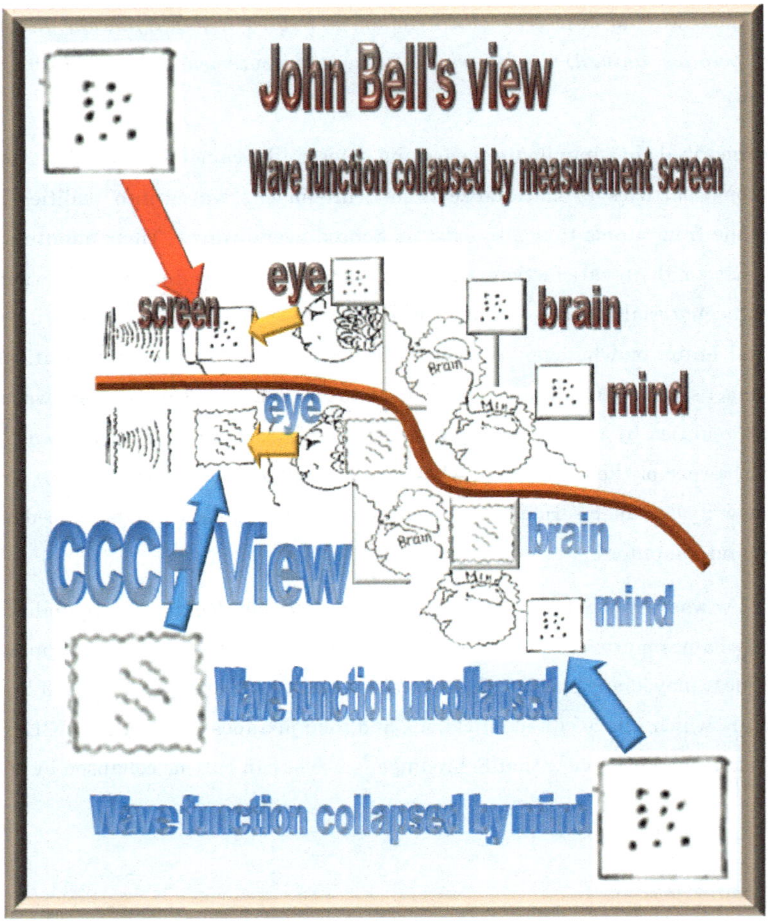

The key features of this diagram are the "squiggles" and "dots:" The squiggles represent the interference pattern of a Schrodinger's wave superposition that hasn't been collapsed into realities. The dots represent objects collapsed from Schrodinger's wave superposition probabilities the "clump" pattern of reality when observed. At what point do the squiggles of Schrodinger's waves collapse into the points of reality?

Bell felt that Schrodinger's waves collapse into realities as soon as they land on the measurement screen, creating "irreversible information." If a person happens to see the measurement screen, as shown in the top portion of Bell's edited diagram, then the information is passed to the eye, the brain, and the mind. The eye, brain, and mind aren't doing anything other than observing the clump pattern of dots created by a particle that became real as soon as it hit the measurement screen.

In Bell's view of our baseball fan analogy, it would be the satellite camera that caused the millions of potential fans to collapse into one real person because when one materialized, a permanent change was made in the atoms of the film, if it is an old-fashioned camera, or electronic memory if it is a modern one. The camera pre-existed the baseball fan. The camera has a history of manufacture with raw materials and labor. It is a real object with a real history. Shouldn't the large object that already exists have priority over the one that doesn't? This view seems so common-sensical that it shouldn't be controversial. The fact that it is still in dispute a hundred years after the formulation of quantum mechanics tells a story that there is more going on than meets the eye.

CCCH mavens believe that Schrodinger's waves do not collapse into dots until information is transmitted through the receptors of the eye into the brain and is analyzed for meaning by the conscious mind. The observing object --- a camera, a Geiger counter, a measurement screen, or a human eye --- is a collection of atoms, so why are their atoms any "better" than the atoms of the object observed?

They believe that Schrodinger's wave superpositions chain up. A small particle in superposition destabilizes a larger object it interacts with, putting it into superposition. The superpositions must end somewhere. Since we never observe items in superposition, perhaps it is our consciousness that causes Schrodinger's wave superpositions to collapse. We see the evidence of superpositions when they create interference patterns, but we never see anything in multiple places and forms at the same time. We only ever see them in one place and in one form, even though we surmise by their interference pattern that they were in superposition before we looked at them.

The idea is that every object in the Universe, from subatomic particles to the Universe itself, exists in Schrodinger's wave superpositions of unresolved possibilities. That includes the atoms inside our brains. The theory is that when a person looks at an object that exists in Schrodinger's waves of superpositions, the superposition images are transferred into our brains by our optic nerves, putting the atoms of our brains in superposition. This theory postulates that only our minds, whose thoughts are "external observers" and not material objects, can cause the Schrodinger's waves of the atoms in our brains to collapse, which then chains down to cause the atoms in the intermediate objects (cameras, Geiger counters, etc.) to collapse, causing the observed object to collapse from superposition to reality.

What is so special about consciousness? Maybe nothing. Maybe consciousness is just a collection of brainwaves of electrical and chemical phenomenon inside the brain, like any other physical process. But perhaps consciousness is superior to other aspects of the Universe. The visible universe is ***deterministic.*** If we know the present properties of a visible object in the universe, we can use mathematics to predict its future properties, according to the laws of physics as they operate over time. Earth's position in its orbit around the sun can be extrapolated far into the future because the laws of motion and gravity are predictable. The unobserved small-particle universe of quantum mechanics is ***probabilistic.*** Until we observe a particle, we don't know where it is or what its properties will be. It doesn't exist yet because it hasn't been observed and measured. But we can still compute the probabilities that when the object is observed it will materialize with a specific location and properties.

Consciousness might be neither deterministic nor probabilistic. There may be no way to mathematically represent human thoughts, no matter how closely we analyze the atomic and subatomic components of neurons in a person's brains. If that is the case, then consciousness is non-computable, and there will never be artificial intelligence capable of thinking as humans do. If consciousness is non-computable, it may be the only entity in the Universe that can collapse mathematically defined probabilistic Schrodinger's waves into mathematically defined deterministic realities. Von Neumann believed that the Schrodinger's wave of a particle can entangle and put into superposition any material object, be it a Geiger counter or a brain, but it cannot entangle the mind. He thought that the mind, as the seat of consciousness, is an entity beyond matter and thus the only entity that can collapse Schrodinger's waves into material realities.

Regarding a radioactive particle encountering a Geiger counter, the CCCH interpretation is that the atom is in a "superposition" state of having emitted and not emitted a radioactive particle. The Geiger counter, being composed of atoms, is therefore also placed in superposition until a human looks at it. Only by hearing the Geiger counter click or an indicator light to flash; and understanding the context of what the click or flash means; does a sentient human being create a reality of an atom decaying.

Erwin Schrodinger conjured up his eponymous "Schrodinger's Cat" to challenge this view that reality is created by observation of a conscious sentient observer:

One can even set up quite ridiculous cases. A cat is penned up in a steel chamber, along with the following device (which must be secured against direct interference by the cat): in a Geiger Counter there is a tiny bit of radioactive substance, so small, that perhaps in the course of the hour one of the atoms decays, but also, with equal probability, perhaps none; if it happens, the counter tube discharges and through a relay releases a hammer that shatters a small flask of hydrocyanic acid. If one has left this entire system to itself for an hour, one would say that the cat still lives if meanwhile no atom has decayed. The first atomic decay would have poisoned it....

the entire system would express this ambiguity of having the radioactivity and the cat in superposition] by having in it the living and dead cat... mixed or smeared out in equal parts.

CCH mavens claim the cat would be alive and dead at the same time, until a human being, with knowledge of what constitutes life and death, looked at it. The person's observation would presumably create the back history that determined whether an atom had decayed and sent a radioactive particle as the harbinger of the cat's demise.

Schrodinger's position is that the cat is a living object, observed by itself. It would presumably know if it was alive and healthy, or about to keel over and die. Even if it was a robotic cat programmed to turn off its lights and fall over if the Geiger counter clicked --- it would still affect atoms in its vicinity. Atoms in the air and on the floor would be disturbed by its robotic "death," perhaps emitting photons that interact with other atoms far away. Wouldn't this start a chain reaction spreading across the universe, thereby conferring a definite life-or-death status on the cat without anyone having to observe it?

This controversy over the possible role of consciousness in bringing objects into reality opens doors that many physicists don't want to walk through. The first forbidden door is theology. Since sentient hominids have only been around for a few million years in a Universe believed to be around 14 billion years old, any invocation of consciousness as a quantum creation process implies that God's consciousness created the Universe. Some contemporary theologians talk about quantum mechanics as being divinely inspired. Most physicists disdain that mixing of science and religion, notwithstanding the fact that many QM founders had an affinity for the philosophy of Eastern religions like Hinduism with its emphasis on reincarnation.

Physicists also do not like the ambiguity of not knowing when Schrodinger's waves collapse. Where is the "cut" between a quantum particle in superposition and visible reality? Is it the measurement screen, the detector, the brain, the mind, or none of the above?

Many Worlds theory avoids these controversies by proposing that Schrodinger's waves neve collapse, but instead spin off infinite separate universes. According to *Many Worlds* mavens, when a Schrodinger's wave interference pattern of a photon or electron impacts a measurement screen, there is never one dot, but quadrillions of dots for each possible point of impact. The entire Universe subdivides into quadrillions of copies, one for each possible impact:

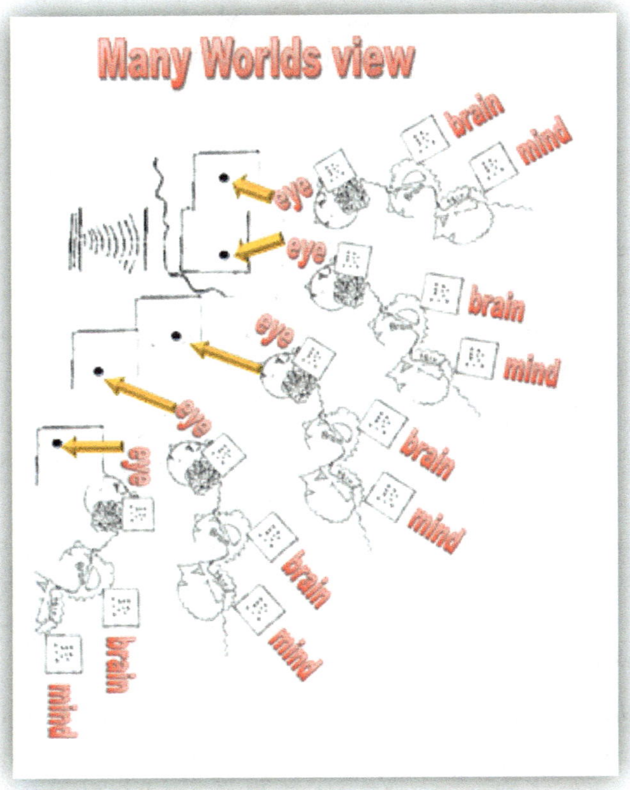

Many Worlds theory has been given a scientific boost in recent years due to the cosmologica theory of "inflation" that proposes that our Universe was created from nothing. *Many Worlds* maven theorize that if it happened once, it could happen again, and again, and again...

Of course, the practical problems are legion. The first is granularity. When the Schrodinger' wave of a particle impacts a measurement screen, there are an infinite number of possible impac points, since the measurement screen can be parsed into at least as many impact points at as ther are atoms on the surface of the measurement screen. There's also a possibility that part of th Schrodinger's wave could be anywhere in the Universe, since Schrodinger's waves are not bound b confined spaces. There is a possibility, however infinitesimal, that the electron or photon coul impact something across the street, on Mars, or on Andromeda, or anywhere. So how do we create an infinity of Universes for a Schrodinger's wave of infinite dimensions? Do we really create Universes that include copies of us, just because we decided to fire one electron through a double slit?

If that's what happens, then every time we look at anything, or decide to do anything, we're spinning off quadrillions of universes. During an average human lifetime, we would create about 10^{100} clones of each of us (and clones of everybody else) in separate universes. There would be almost infinitely more copies of you and me in the "Multiverse" than there are subatomic parties in the whole of our known Universe. But that's not all. Every one of those staggering numbers of universes would create infinite universes of their own every time an event happened in them. And it wouldn't just be our decisions and actions that created these other universes. Every subatomic particle in the universe whose Schrodinger's waves exist in superpositions would be spinning off universes. Can an infinity of universes spawn infinite universes that spawn infinite universes that spawn more infinite universes? *Many Worlds* mavens answer that question affirmatively:

Every quantum transition taking place in every star, in every galaxy, in every remote corner of the universe is splitting our local world on Earth into myriads of copies. Here is Schizophrenia with a vengeance – Bryce DeWitt.

A variation on the theory is that the Universe isn't multiplying, but rather that an infinitely thick universe is subdividing itself into thinner and thinner slices. But it comes down to the same thing of whether our Universe spins off infinite junior league universes, or an infinite universe subdivides into quadrillions of universes...that subdivide into quadrillions of universes...that subdivide....

It has also been conjectured that the "*many worlds* schizophrenia" might be limited to our minds. Perhaps when we observe events, our minds create separate realities for every possible outcome, and every one of our myriad "minds" sees a separate universe, even if these universes don't have any hard reality behind them. This is known as the "Many Minds" theory. But wouldn't we need separate brains to house each of those separate minds? And if we have separate brains, then we'd have to have separate bodies, and again we are back to the problem of creating quintillions of quintillions of quintillions of universes.

Perhaps Schrodinger's waves, at least for objects large enough to be visible, always collapse spontaneously, because visible objects are self-observing. Any particle with an electric charge is observed by other charged particles. Any particle with mass exerts a gravitational force. Since quintillions of subatomic particles are packed into visible objects, they may collapse each other's Schrodinger's waves spontaneously. Large bodies have gravitational fields that theoretically extend across the universe. Perhaps all bodies in the universe large enough to exert or feel the tug of gravity will have their Schrodinger's waves collapsed automatically. This process of self-collapse has been termed "decoherence" based on the idea that Schrodinger's waves of mathematical probabilities lose their coherence due to frictions resulting from encounters with particles near them, and from the far-reaching effects of electrical and gravitational forces.

If that is the case, then perhaps only particles too tiny to exert or be influenced by electric and gravitational charges will not have definite realities until we observe them. However, if there is any such dividing line between the microscopic and macroscopic worlds, we don't know where the line is. As our ability to isolate objects from external influences increases, we are observing interference patterns made by ever-larger objects whose Schrodinger's waves haven't collapsed into realities. We have observed Schrodinger's wave interference patterns for objects containing several hundred molecules.

We don't know whether reality chains begin with large objects and work their way downward until a certain point of smallness is reached where reality goes away; or whether Schrodinger's waves start with microscopic objects and work their way upward until reality spontaneously coalesces. We don't know whether a camera, Geiger counter, or human eye collapses the Schrodinger's wave of a subatomic particle by the mechanical process of detecting its photons or ionization trail; or whether the subatomic particle entangles the detecting devices in its Schrodinger's wave of probabilities until a human consciousness hears the "click" of the Geiger counter or sees its screen light up.

Other quantum interpretations don't have anything at all to do with consciousness. However, these other theories have unresolved inconsistencies and / or missing components that make them incomplete.

The interpretation known as Bohmian Mechanics theorizes that Schrodinger's waves have "hidden realities" that we can't detect. Perhaps these hidden realities give the particle a definite set of deterministic properties that only appear to be a probability Schrodinger's wave because we don't know how to observe the hidden parts of reality (maybe an extra dimension we don't know about?) until we lay eyes on the particle and observe as much of it as our three-dimensional eyes can perceive. This is also known as "Pilot Wave Theory" because of its idea that the hidden variables create a pilot wave that pushes the particle along a predetermined path that only appears to be probabilistic because we don't see the hidden variables. But if we can't see them, how do we know they're there?

Or perhaps everything in the Universe that will ever happen was predetermined with the Big Bang. We think we are making decisions of free will to decide when and where to observe particles, but that decision may have been predetermined 14 billion years ago. Everything in the

Universe that was, is, or will be, always existed, exists, will exist in a predetermined state that only appears to be a Schrodinger's wave before we see it.

Or perhaps Schrodinger's waves never collapse, and everything in the Universe is always a superposition of possibilities. Perhaps Schrodinger's waves only appear to collapse because evolution has programmed our brains to winnow down superpositions into a single version of reality that biological brains of finite neural capacity are capable of comprehending. As British author Aldous Huxley wrote in ***The Doors of Perception:***

The suggestion is that the function of the brain and nervous system and sense organs is in the main eliminative and not productive. Each person is at each moment capable of remembering all that has ever happened to him and of perceiving everything that is happening everywhere in the universe. The function of the brain and nervous system is to protect us from being overwhelmed and confused by this mass of largely useless and irrelevant knowledge, by shutting out most of what we should otherwise perceive or remember at any moment, and leaving only that very small and special selection which is likely to be practically useful.... To make biological survival possible,

Huxley, Aldous. The Doors of Perception and Heaven and Hell (pp. 8-11). HarperCollins. Kindle Edition.

We are only beginning to scratch the surface of quantum controversies. The next one is the question: "When a Schrodinger's wave is collapsed into reality, does the collapse travel backward in time and space as well as forward?"

When we see an object, we usually infer the history of where it was before we saw it. But if we believe that the object didn't exist until it was observed, how could its history have existed before then? Does the reality chain travel backward in time and space to the point where the particle was emitted from its source? There is an interpretation of quantum mechanics called "transactional" theorizing that a particle's past reality materializes when it is observed in the present. For this to happen, the "pilot waves" theorized by Bohmian Mechanics would have to travel backward in time and space as well as forward.

The most extreme "consciousness creates reality" proponents theorize that reality chains extend from the present backward to the beginning of the Universe. For example, a photon that originated millions or billions of light years ago in a distant galaxy travelled across the Universe to reach our eyes and enter our minds. The champions of consciousness say that we not only create the photon when we see it, but also materialize its reality backward in time through all those millions or millions of years, all the way back to its instant of origin in a distant part of the Universe. Physicist John Wheeler said:

"We are participators in bringing into being not only the near and here but the far away and long ago. We are in this sense, participators in bringing about something of the universe in the distant past and if we have one explanation for what's happening in the distant past why should we need more?"

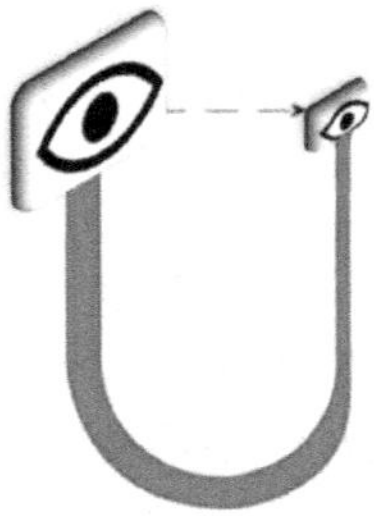

This recalls the old Chinese proverb of whether a tree falling in a forest makes any noise if there is no one within earshot to hear it. According to Wheeler's theory of retro-causality, an observation of a falling tree creates the entire backward reality. It causes the tree to germinate from a seed that became the tree that had some structural deficiency that caused it to fall when you happened to stop by and witness it. Your observation of it falling created its entire proto history, all the way back to the Big Bang when a certain set of atoms fell into a pattern that would emerge into the form of a tree billions of years later. The consciousness of all human beings viewing the Universe at this moment created its back history going back to its origin.

Who, then, created history during all those billions of years before humans evolved consciousness? Was there a super-consciousness that preceded the Universe, and decreed, "Let there be light?" Or is everything in the Universe happening all at once, and separation of events by time an illusion? Or was the Universe in some lower state of order until we acquired the intelligence to watch it and discern its meaning? We don't have to answer these questions to perform our experiment. All we need to assume is that consciousness can store information about reality, which we know it does whenever we remember something.

If there is any conclusion we can draw at this point, it is that the engine of quantum mechanics remains obscure, making it as yet impossible to confirm or refute any of the interpretations. Quantum physicist David Bohm, who devised the "Bohmian Mechanics" interpretation of quantum theory, held the suitably ambiguous view that:

...quantum theory itself can be understood without bringing in consciousness, and that as far as research in physics is concerned, at least in the present general period, this is probably the best approach. However, the intuition that consciousness and quantum theory are in some sense related seems to be a good one.

The Quantum Cockroach

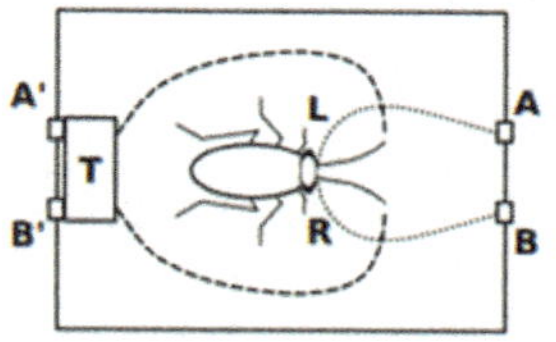

Fig. 4 Proposed experimental setup. A photon impinges on A or B, and an optical fiber, represented by the dotted line, takes it to either the left (L) or right (R) eye, respectively. If a photon reaches L, the cockroach is conditioned to push a button at the end of a circuit (dashed line), and if the L button is pushed, a single photon is emitted at a precise and very short window of time.

We've taken a quantum journey to the ends of our Universe, and maybe a quadrillion others in the Multiverse. Let's take the opposite tack of looking at some quantum microcosms. For example, can a cockroach collapse a Schrodinger's wave? Don't think this question is too weird for quantum mechanics, because it has already been asked:

https://arxiv.org/pdf/1609.00614.pdf

Can we Falsify the Consciousness-Causes-Collapse

Hypothesis in Quantum Mechanics?

J. Acacio de Barros · Gary Oas

Since we want to test CCCH [Consciousness Causes Collapse Hypothesis], like Yu and Nikolic, we start with the eyes as photodetectors. Nature provides us with exceptionally good photo-detectors in the kingdom of Animalia.... Perhaps one of the best candidates for such conditioning experiments is the cockroach.... ***For our purpose, we will also assume that the cockroach is a conscious being.***

This is, of course, a controversial assumption, but the alternative would be to do our proposed experiment with more complex animals (say, humans). However, as it will become clear below, this assumption will not invalidate our conclusions, as they will apply to any animal. The idealized experiment we propose is simple and does not rely on entangled states (as does Kim et al.'s). Imagine we have a cockroach who has been conditioned to respond to single photons in the following way [the quantum equation for a cockroach]:

where

$$\rho^{\text{roach}}_{\text{ready}} = |\text{cockroach ready}\rangle\langle\text{cockroach ready}|,$$

$$\rho^{\text{box}}_{\text{ready}} = |\text{box ready}\rangle\langle\text{box ready}|.$$

and

$$\rho'_{0,0} = |0_{A'}, 0_{B'}\rangle\langle 0_{A'}, 0_{B'}|.$$

This system would evolve the following way:

$$\begin{gathered}
\rho_{1,0} \otimes \rho^{\text{roach}}_{\text{ready}} \otimes \rho^{\text{box}}_{\text{ready}} \otimes \rho'_{0,0} \rightarrow \\
\rho_{0,0} \otimes \rho^{\text{roach}}_{\text{left antennae}} \otimes \rho^{\text{box}}_{\text{ready}} \otimes \rho'_{0,0} \rightarrow \\
\rho_{0,0} \otimes \rho^{\text{roach}}_{\text{ready}} \otimes \rho^{\text{box}}_{\text{gen.photon A'}} \otimes \rho'_{0,0} \rightarrow \\
\rho_{0,0} \otimes \rho^{\text{roach}}_{\text{ready}} \otimes \rho^{\text{box}}_{\text{ready}} \otimes \rho'_{1,0},
\end{gathered}$$

where the label for the states should make them evident. A similar evolution would happen to $\rho_{0,1}$, leading to

$$\begin{gathered}
\rho_{0,1} \otimes \rho^{\text{roach}}_{\text{ready}} \otimes \rho^{\text{box}}_{\text{ready}} \otimes \rho'_{0,0} \rightarrow \\
\rho_{0,0} \otimes \rho^{\text{roach}}_{\text{ready}} \otimes \rho^{\text{box}}_{\text{ready}} \otimes \rho'_{0,1}.
\end{gathered}$$

Finally, if we started with a superposition given by, say, the state

$$\rho_{1,1} = \frac{1}{2}\left(|0_A, 1_B\rangle\langle 0_A, 1_B| + |1_A, 0_B\rangle\langle 0_A, 1_B| + |0_A, 1_B\rangle\langle 1_A, 0_B| + |1_A, 0_B\rangle\langle 1_A, 0_B|\right).$$

we would end with the linear evolution

$$\begin{gathered}
\rho_{1,1} \otimes \rho^{\text{roach}}_{\text{ready}} \otimes \rho^{\text{box}}_{\text{ready}} \otimes \rho'_{0,0} \rightarrow \\
\rho_{0,0} \otimes \rho^{\text{roach}}_{\text{ready}} \otimes \rho^{\text{box}}_{\text{ready}} \otimes \rho'_{1,1}.
\end{gathered}$$

Clearly, if the experiment could be performed like above, if the input is a superposition, we can take the partial trace over all other variables, and the output will also be a superposition. In other words, because the evolution is linear, the partial trace over $H_p \otimes H_c \otimes H_b$ of $\rho_{0,0} \otimes \rho^{\text{roach}}_{\text{ready}} \otimes \rho^{\text{box}}_{\text{ready}} \otimes \rho'_{1,1}$ would result in $\rho'_{1,1} \in H_{p'}$. However, if the cockroach's mind causes a collapse of the wave function inside the box, then the dynamics would not be linear, and the output would be the proper mixture

$$\rho'_{\text{mixture}} = \frac{1}{2}\left(|1_{A'}, 0_{B'}\rangle\langle 1_{A'}, 0_{B'}| + |0_{A'}, 1_{B'}\rangle\langle 0_{A'}, 1_{B'}|\right),$$

and not the pure state $\rho'_{1,1}$.

Since we want to test CCCH, like Yu and Nikolic, we start with the eyes [of the cockroach] a photodetectors.

Our eyes are complex structures that enhance our vision beyond what reality presents. The sharpen the edges of objects of different shapes, colors, and distances to differentiate them --- a evolutionary advantage that helps detect camouflaged predators by increasing the contrast betwee them and their backgrounds. So perhaps the structure of the eye has evolved to collaps Schrodinger's waves whenever photons are seen. Or perhaps the neural networks in our brains ar collapsing Schrodinger's waves electrochemically, without consciousness being involved.

If biology ***is*** involved in collapsing Schrodinger's waves into realities, how far down the "foo chain" does an organism's brain collapse them? The following article seeks to determine whether Euglena (single-celled organism with a rudimentary "eyespot") can do it:

https://arxiv.org/ftp/quant-ph/papers/0509/0509042.pdf

Does consciousness really collapse the wave function?

A possible objective biophysical resolution of the measurement problem

Fred H. Thaheld* 99 Cable Circle #20 Folsom, Calif. 95630 USA

3. Can Euglena collapse the wave function?

I would now like to address the issue as to whether humans alone can collapse the wave function, since there is such a wide range of both vertebrates and invertebrates, all of whom preceded us on this planet… I have chosen the Euglena gracilis, a unicellular protozoan…

The question now before us is: Are Euglena capable of collapsing the wave function of superposed states, either as an animal with its eyespot or as a plant with its chloroplast, in a manner analogous to the human eye? We will not address the matter of whether they can interpret the collapsed information or their degree of self-consciousness.

Many Worlds maven David Deutsch suggested using a frog to find all its counterpart frogs in the other universes of the Multiverse:

At the beginning of the experiment, the tangible frog will have a large set of shadow counterparts, all initially alike. But shortly afterwards they will no longer all be alike. Any particular one of them is unlikely to see a photon immediately. But what is a rare event in any one universe is a common event in the multiverse as a whole. At any instant, somewhere in the multiverse, there are a few universes in which one of the photons is currently striking the retina of the frog in that universe. And that frog jumps.

He says he is certain that many copies of himself are out there along with the quantum frog:

While I was writing that, hosts of shadow Davids were writing it too. They too drew a distinction between tangible and shadow photons; but the photons they called 'shadow' include the ones I called 'tangible', and the photons they called 'tangible' are among those I called 'shadow'.

Deutsch, David. The Fabric of Reality (p. 53). Penguin Publishing Group. Kindle Edition.

Albert Einstein asked, tongue in cheek, if Schrodinger's waves could be collapsed by a mouse. Physicist John Stewart Bell asked if you had to have a Ph.D. in Physics to know how to collapse a Schrodinger's wave. We've talked about the quantum universe, now we're down to quantum euglenas, cockroaches, and frogs. Talk about "playing with quantum guns!"

Can we narrow the controversies of Quantum Mechanics by proving that consciousness ***isn't*** required to collapse Schrodinger's waves into realities? An experiment purporting to do this was reported a few years ago:

http://www.danko-nikolic.com/wp-content/uploads/2011/10/Yu-and-Nikolic-Qm-and-consciousness-Annalen-Physik.pdf

Quantum mechanics needs no consciousness

Shan Yu and Danko Nikolic

The nature of human consciousness and its relation to the physical reality is one of the most puzzling fundamental issues about ourselves and about our interaction with the world we live in... Specifically, it has been suggested that the consciousness plays a vital role for the quantum mechanics as it is capable of, and even required for affecting physical events through the collapse of the wave function.

[John] Von Neumann was among the first people to suggest a unique relation between the human observer and the measurement problem...

In conclusion, the available evidence does not indicate that the observer's explicit phenomenal representation about the outcome of a measurement plays a role in collapsing the wave function. Thus, the idea that by mere observation the experimenter creates physical reality does not seem viable.

However, this experiment was soon disputed:

https://arxiv.org/ftp/arxiv/papers/1707/1707.01346.pdf

Comment on the paper "Quantum mechanics needs no consciousness", by Yu and Nikolic (2011)

Catherine M Reason

This is a brief comment on the paper ***"Quantum mechanics needs no consciousness"*** *by Shan Yu and Danko Nikolic.*

Yu and Nikolic argue that the "consciousness causes collapse hypothesis" interpretation of quantum mechanics, or CCCH, can be falsified by a particular experimental setup. This claim is incorrect, and the cause of the error appears to be a confusion over where and when a collapse can be assumed to occur.

.... The claims made by Yu and Nikolic, that the CCCH [Consciousness Causes Collapse Hypothesis] predicts a second-order interference pattern at D0 whenever "which-path" information is available but not consciously observed, are therefore unfounded. ***Their claim to have falsified the CCCH is therefore incorrect.***

Another refutation came from the cockroach experimenters:

http://online.sfsu.edu/barros/publications/page9/files/deBarrosOas2017.pdf

Can We Falsify the Consciousness-Causes-Collapse Hypothesis in Quantum Mechanics?

J. Acacio de Barros · Gary Oas

Received: 6 September 2016 / Accepted: 24 June 2017

One of the central issues within Quantum Mechanics (QM) is the measurement problem.... Among the proposed solutions, perhaps one of the most controversial is von Neumann's idea that a measurement is the result of the interaction of a (conscious) mind with matter.... To understand Kim et al.'s experiment, it is important to notice first that it is a fourth-order interference experiment......We saw that their proposal had a fatal flaw, as it did not consider the fact that to observe fourth-order interference requires coincidence counting...... Therefore, any experiment trying to falsify CCCH [Consciousness Causes Collapse Hypothesis] ... is doomed.

This is common in quantum mechanics. A controversy that has agitated the minds of quantum physicists since the days of Einstein and Bohr is said to have been settled by experimental results. Then other quantum physicists say they found a flaw in the original experimenter's method. We must be wary of sensationalism when "quantum discoveries" are reported in the popular press. For now, we continue to be unsure of whether Schrodinger's wave collapses spontaneously due to inanimate physical processes, or whether it requires the biological processes of a euglena, a cockroach, a mouse, or even a physics student.

Quantum Immortality

Immortality is defined as the retention of a person's memory after brain death. Let's recall the theory of Roger Penrose that the biological retention of information in the brain is reposed in a quantum soul that survives brain death:

https://www.express.co.uk/news/science/728897/LIFE-AFTER-DEATH-consciousness-continue-SOUL

LIFE AFTER DEATH: Shock claim of evidence showing consciousness may continue as a SOUL

...According to some well-respected scientists, quantum mechanics allows consciousness to live on following the body's eventual demise.

.... British physicist Sir Roger Penrose agrees and believes he and his team have found evidence that protein-based microtubules – a structural component of human cells – carry quantum information – information stored at a sub-atomic level.

Sir Roger states if a person temporarily dies, this quantum information is released from the microtubules and into the universe.

Sir Roger added: "If they're not revived, and the patient dies, it's possible that this quantum information can exist outside the body, perhaps indefinitely, as a soul."

There's also a theory of quantum immortality derived from the *Many Worlds* interpretation. It postulates myriads of copies of us in the Multiverse, some that never die. Thus, our memories might live in multiple copies of ourselves in other universes. Popular physics author Dr. Sean Carroll writes:

I will focus on the approach I feel is clearly the most promising route, the Everett or Many-Worlds formulation of quantum mechanics. Many-Worlds has been enthusiastically embraced by many physicists, but it has a sketchy reputation among people who are put off by a proliferation of other realities containing copies of themselves.

...In fact, I'll make a confession: there really are such worlds.... According to Many-Worlds, when I generated that random number, the universe split into 2^50 [2 raised to the 50th power] copies (that's 1,125,899,906,842,624, or approximately 1 quadrillion), each of which carries a slightly different number.

The situation in Many-Worlds quantum mechanics is exactly the same way, except that now more than one person can descend from a single previous person....Every one of those people has a reasonable claim to being "you." None of them is wrong. Each of them is a separate person, all of whom trace their beginnings back to the same person.

In Many-Worlds, the lifespan of a person should be thought of as a branching tree, with multiple individuals at any one time, rather than as a single trajectory—much like a splitting amoeba.... The world duplicates, and everything within the world goes along with it.

Carroll, Sean. Something Deeply Hidden (pp. 134-140). Penguin Publishing Group. Kindle Edition.

Dr. Hugh Everett, progenitor of the *Many Worlds* Theory, was a chain-smoking, morbidly obese alcoholic --- but otherwise renowned physicist, mathematician, and Wall Street stock guru --- who died in his early 50s. Perhaps he presumed that in some other universe there was a copy of him that didn't smoke, drink, or overeat. Before he died, he said he wanted to be thrown out with the garbage, perhaps because he thought he had billions of other Multiverse bodies to fall back on.

His reckless disregard for his life may have induced his daughter to commit suicide. Her suicide note indicated that she wanted to have her ashes thrown out with the garbage, to "end up in the correct parallel universe to meet up with Daddy." However, Dr. Everett's son Mark Oliver Everett became the rock superstar "E" in the band Eels, so perhaps he inherited the musically gifted side of his father's personality. At least in that non-sentient way Dr. Everett passed on genetic information to one of his surviving children, although his dubious parenting may have short-circuited his other child. Dr. Everett's son wrote a mellow song, referring to his dad's scientific outlook, and the melancholy demise of his sister:

https://www.youtube.com/watch?v=0TfqbuTBqX8

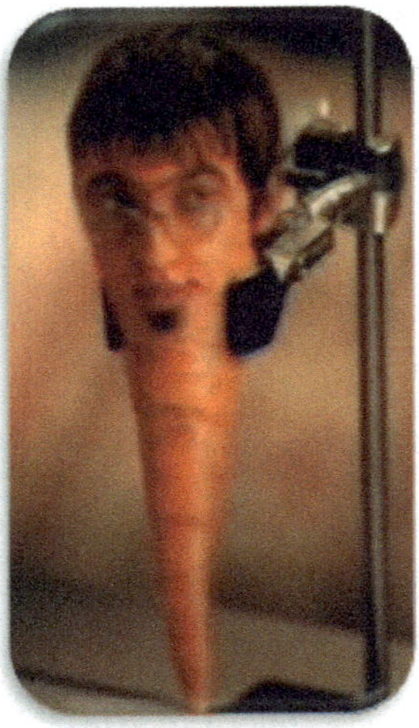

John Von Neumann, a lifelong agnostic, went to the other extreme of fearing death so muc that he converted to Catholicism when he contracted fatal cancer at 53. He referred to Pascal' Wager just before his death: "There probably has to be a God. Many things are easier to explain i there is than if there isn't."

Most of us have a bit of Dr. Everett's devil-may-care attitude combined with Dr. vo Neumann's terrifying fear of death when it becomes imminent. Pascal's Wager remains the sam now as 370 years ago --- try to live the most moral life you can and hope for the best when you die *Many Worlds* maven David Deutsch, an Oxford University physicist, added a multiple-choice mora tone to the *Many Worlds* theory:

"In the many-worlds interpretation, when you make a choice, the other choices also happen. I there is a small chance of an adverse consequence, say someone being killed, it seems on the face of i that we have to take into account the fact that in reality someone will be killed, if only in anothe universe."

Perhaps Erwin Schrodinger posed the question of immortality the way most "hopefu agnostics" feel about it. Like Charles Darwin and Blaise Pascal, he knew there was no answer tha the science of his time could provide to the question of the Creator and the immortal soul. He als knew that it was futile to expect science to refute it:

Some of us, in particular healthy and happy youth, succeed in shoving them [these question of immortality] aside for long periods; others, in advanced age, have satisfied themselves that there i no answer and have resigned themselves to giving up looking for one, while others again are haunte [by] the questions concerned with the 'other world', with 'life after death', and all that is connected with them.

Whether or not our essence lives on in an immortal soul, we do know that we pass information about ourselves in non-sentient ways. Our physical form is encoded in our genes. In addition to genes sculpted by natural selection, there is also a process called "epigenesis" whereby some of life's acquired characteristics are imprinted on the sperm and egg cells before being transmitted to our offspring. Parents who suffer famine as children may have their DNA and/or non-DNA cell components marked with acquired food-minimization characteristics passed on to their children. Instead of requiring dozens of generations of natural selection to breed children whose bodies conserve calories, the transition is made in one generation.

Does natural selection and / or epigenesis encode behaviors, such as the innate fear of snakes into the DNA of animals in places where poisonous snakes are encountered? Animals born with randomly selected reflexes to avoid poisonous snakes would be expected to live longer, and therefore create more offspring than animals that don't. So, it could be the natural selection of animals whose nervous systems are randomly wired to flee slithering things that causes snake-averse critters to prevail in the gene pool. If an animal is poisoned by a snake but survives, it is possible it transmits that information to its offspring by epigenesis. However encoded, the instinct is complex. How does a spider "know" how to build its web, or a conch know how to build its beautiful shell, or a salmon know to swim the oceans then return to the river of its birth to spawn, or a bird how to build a nest?

Can this information really be stored in DNA and other cell structures? We might be able to eventually parse DNA to account for the physical forms of organisms, but will we ever be able to parse DNA for the information stored as innate instincts that organisms are born with? What part of the DNA tells the conch how to build its shell? Could there be a "quantum memory" that stores information about each species outside its material body? Biologist Rupert Sheldrake asserts that quantum physics "overturns" the idea that instinctive memories are transmitted genetically.

If memories are stored in individual animals' brains, then anything an animal learns is confined to its own brain. When it dies the memory is extinguished. But if memory is a resonant phenomenon through which organisms specifically resonate with themselves in the past, individual memory and collective memory are different aspects of the same phenomenon; they differ in degree, not in kind.

This hypothesis is testable. If rats learn a new trick in one place, then rats all over the world should be able to learn the same trick quicker. The more rats that learn it, the easier it should become everywhere else. There is already evidence from one of the longest series of experiments in the history of psychology that rats do indeed seem to learn quicker what other rats have already learned. The more that learned to escape from a water maze, the easier it became for others to do so.

Likewise, humans should be able to learn more easily what others have already learned. New skills like snowboarding and playing computer games should become easier to learn, on average....

Sheldrake, Rupert. Science Set Free (pp. 207-208). Potter/Ten Speed/Harmony/Rodale. Kindle Edition.

This is a dubious theory, insofar as rats go, because after thousands of years living among humans, mice still haven't learned to avoid mouse traps. However, some historians have commented on how rapidly knowledge spreads among humanity. Ideas for the design of new machines, new theories of mathematics, and so on, seem to develop in many countries simultaneously. The most likely explanation is mundane. In the fields of industrial and scientific knowledge, the world is small. Engineers and scientists everywhere read the same journals. People with knowledge in one country are hired by governments of other countries who want to obtain it. Technologies that seem new took decades to incubate. Knowledge was already widespread before technology became visible in many places simultaneously. Those are the likely, logical reasons for the rapid spread of knowledge among humanity. But what if there is some quantum sphere of knowledge that becomes known to all of mankind as soon as one person discovers it?

Another question is whether evolution is directed forward by a higher intelligence, the "intelligent design" hypothesis. The conch shell seems like it might have been designed as an intelligent work of art, but it is more likely merely an organism's non-sentient response to billions of years of random pounding by surf and sand. Or perhaps there is a quantum store of "resonance" knowledge that builds upon itself in each species as Sheldrake believes, thereby driving the species forward to higher states of development.

Sheldrake takes his concept of universal quantum knowledge into the Hereafter:

The resonance theory of memory also opens up a religious question. All religions take it for granted that some aspect of a person's memory survives that person's bodily death. In Hindu and Buddhist theories of reincarnation or rebirth, memories, habits or tendencies are carried over from one life to another.

This transfer of memory is part of the action of karma, a kind of causation across time; actions bring about effects in the future, even in later lives. In Christianity there are several different theories of survival, but all imply a survival of memory... Some Protestants believe that after death everyone goes to sleep, only to be resurrected just before the Last Judgment. But this theory too requires a survival of memory because the Last Judgment would be meaningless if the person being judged had forgotten who he was and what he had done.

Sheldrake, Rupert. Science Set Free (pp. 207-208). Potter/Ten Speed/Harmony/Rodale. Kindle Edition.

Sheldrake's views are pseudo-science, though some are taken seriously by mainstream quantum physicist David Bohm. Many quantum mechanics founders from the early-to-mid 20th Century, and many quantum scientists today, expressed affinity for Eastern religions that postulate a transcendent knowledge residing outside our brains.

This affinity for the metaphysical may derive from the conflict between physics and biology. Physics, according to the Second Law of Thermodynamics, says that the Universe must decay into random chaos. Biology says that life must evolve toward higher orders of complexity and intelligence. Quantum mechanics may bridge the gap by explaining how information acquired by observation and experience is used by living organisms to reverse nature's entropy from chaos to order.

What about the sentient information stored in our minds? Does the brain really have enough neurons to store every sight, sound, smell, and emotion we have ever heard in our lives? How are we able to think about our thoughts --- for example remembering a thought we had years or decades ago and refining it by comparing it to the myriad thoughts that have come into our minds since then.

How are our minds able to discern the physical and mathematical structure of the Universe, from the tiniest subatomic particles to the largest trans-galactic structures of matter, energy, time, and space? Our minds seem curiously well-tuned to understanding what makes the Universe tick.

Are there quantum processes that multiply our mental capacity beyond what neural connections alone permit? Perhaps the atoms in our brains are in quantum superpositions like reality itself. Our emergent knowledge of quantum computing leads us to believe that building computers with quantum superposition in each of their "qubits" (quantum bits) will allow almost infinite amounts of information to be stored in machine memory. Perhaps our brains work like that. Or perhaps there are higher dimensions of quantum knowledge in which the instincts and knowledge acquired by humanity are reposed.

Let's reconsider Aldous Huxley's view:

Each person is at each moment capable of remembering all that has ever happened to him and of perceiving everything that is happening everywhere in the universe. The function of the brain and nervous system is to protect us from being overwhelmed and confused by this mass of largely useless and irrelevant knowledge ....

If the brain is dissolved by death, then maybe that's the time when we really start to know everything.

Quantum Erasure

We're beginning to understand that quantum mechanics describes a new component of the Universe, previously known to consist of mass, energy, space, and time. The new quantum component is information. When we observe an object, we create information about it. If the object is a subatomic particle, the information we create by observing it appears to bring the particle into existence.

To conduct our soul-searching experiment, we need to know whether a particle's reality exists only so long as the information about that particle remains in a person's brain. This assumes that the information is only in the person's brain and has not been copied to digital memory or an analog medium like photographic film. If the information resides only in a person's brain, and that person dies, does the reality created by that information go away?

This question was posed to NASA physicist Dr. Tom Campbell:

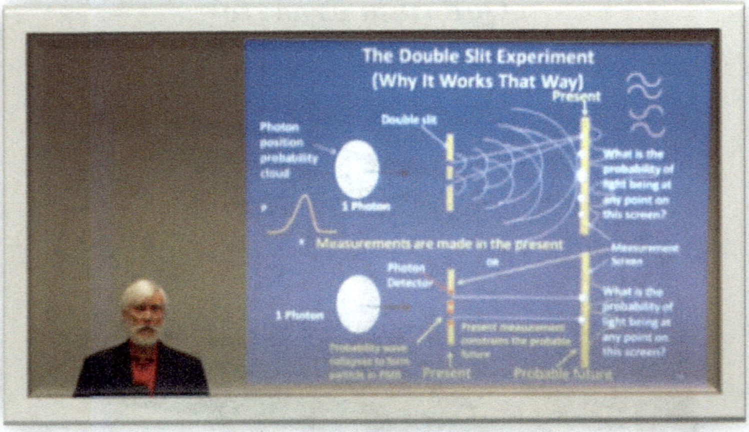

https://youtu.be/BhMIz_iJtzQ?t=2353

"What happens if you get information, and so, it's here?" asks a lady in the audience. "And then what happens if everybody who knows that information dies? Does the information remain?"

"Nope," answers Campbell. "The information's ***gone.*** *If the information's gone, it can change. So, let's say that person with his telescope took that measurement and there's a nebula there and he takes a picture of it and other people come in and they take a picture of it. But let's say after ten people took the picture and they all saw the same thing.*

.... let's say that area gets a disease and they all fall over dead and all the pictures burn in the fire, etc., then the information's gone. No information at all what's in that part of space. Now sometime later, somebody else gets a telescope and they look up there, they take a random draw out of all the possibilities and that's what they get, and it doesn't have to be the same thing."

"Wow!' *exclaims someone in the audience.*

"You see," continues Campbell, "because there's no information there. It's the same as the quantum eraser experiment, right? That's a delayed eraser experiment. So now they didn't see that nebula. They saw a red dwarf [star] or something, because that was another one of the possibilities. And now everybody who looks after that will see a red dwarf. Information comes and goes. And when it goes, it's gone. It's not the consciousness. It's not the measurement. It's the ***information*** *that's the key here. Because this reality is based on information."*

Dr. Campbell concludes that information is elided from the Universe when the people who have that information in their brains die. Of course, it's an extreme view to suppose that stars and galaxies materialize when someone looks at them for the first time and disappear if that person dies without recording the information on a permanent memory device. As extreme as that view might be, it is nothing compared to *Many Worlds*, which purports to believe that we create a whole new universe... or two, or three, or quadrillions, every time we sneeze.

If our souls are immortal, then we are hoping to prove the opposite of Dr. Campbell's theory --- to prove that the information in our consciousness is ***not*** elided from the Universe when we die. To test for immortality, we want to put a piece of information in a person's brain, and nowhere else, and then be able to infer whether than information goes away when the person is deceased.

We will be performing a type of experiment known as "quantum erasure." These experiments test what happens to particles when information about them is elided. The particle is first brought into existence by collapsing its Schrodinger's wave into reality by a detector, then after an interval of time is winked back into non-reality by having the detector's information obscured or elided.

Reputable physicists want to make sure their experiments are "pure quantum" in the sense of not being contaminated by objects that might introduce non-quantum phenomenon that collapses the Schrodinger's waves of particles into realities before they are collapsed by the apparatus in the experiment. The usual "quantum eraser" is a complex apparatus like the one shown below, consisting of prisms, mirrors, half-mirrors, and "coincidence counters:"

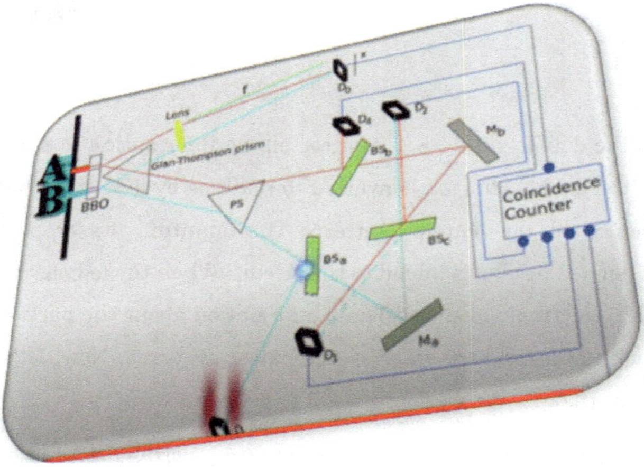

The PBS ***Space Time*** series has an interesting video showing how these experiments are run:

https://www.youtube.com/watch?v=8ORLN_KwAgs

The schematic shown below shows the principle of these quantum eraser experiments:

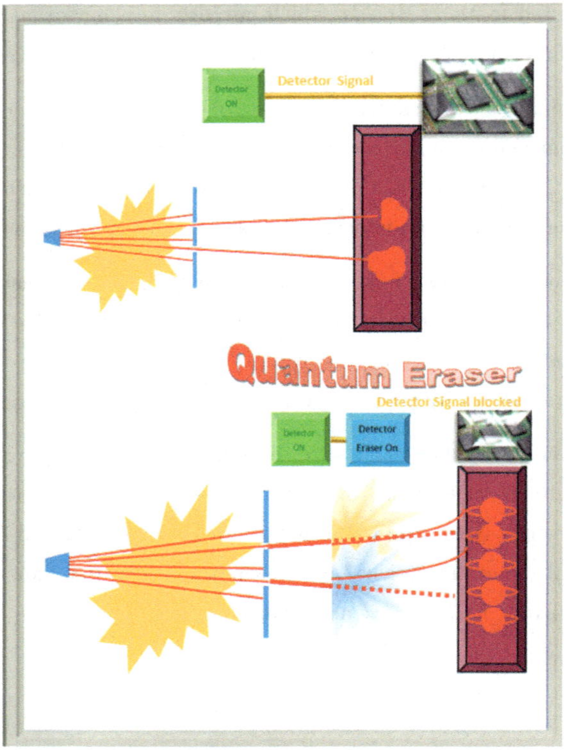

It shows the double slit with a detector piping information to a computer memory. Particles passing through the double slits are converted to realities by detection that makes them land on the measurement screen in the "clump" pattern. The quantum eraser portion is at the bottom. A "detector eraser" shown in blue is added to the circuit. When turned on, it blocks transmission of the detector's signal to electronic memory. Thus, information about the particle is lost, its Schrodinger's wave springs back to life, and the interference pattern returns. Now, let's take an inspiration from Werner Heisenberg that:

The observer has, rather, only the function of registering decisions, i.e., processes in space and time, and it does not matter whether the observer is an apparatus or a human being…

Let's introduce a human observer into the circuit between the detector and the eraser:

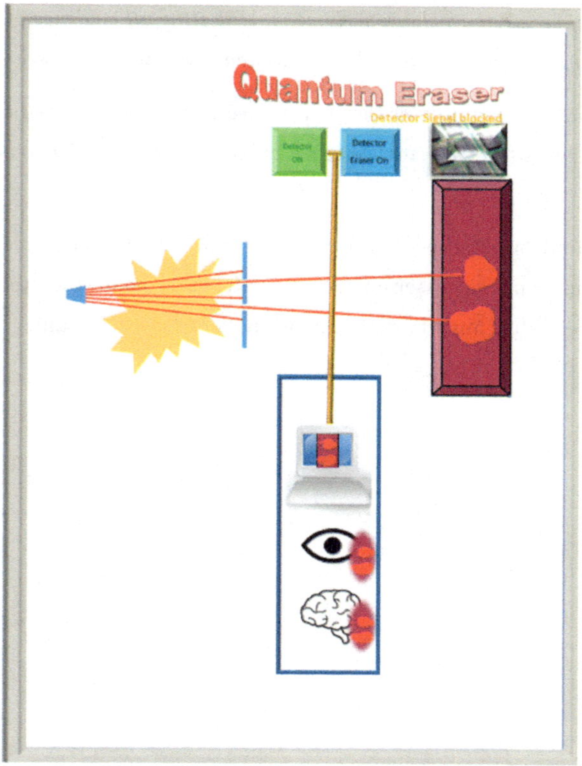

Before going to the eraser, the signal will be output to a TV screen the same as on the measurement screen. Each particle passing the slits will land in a clump. A person will be watching the screen. The "which-path" information will pass through the person's eyes and into the brain. Even if the information is erased before it gets to the computer memory, we should still see a clump pattern on the purple measurement screen because a person's brain is a memory device that stores the which-path information the same a computer memory would. We want to prove that if a human being looks at the information visually, then the Schrodinger's wave cannot be restored by erasing the which-path information that would have gone to the computer memory.

To prepare for this experiment, I suggest watching Tom Campbell's video on quantum erasure from the beginning:

https://youtu.be/BhMIz_iJtzQ

Then this video by Eugene Khutoryansky, explaining the effect of information in the brain during quantum erasure.

https://youtu.be/SzAQ36b9dzs

For a quantum eraser to work properly, it must destroy all evidence of which of the two holes the particle has passed through. This includes any evidence existing anywhere in the Universe.

This means that the quantum eraser will not work if we have looked at the detectors prior to erasing the measurement, since the evidence of which hole the particle has passed through ***will still exist in our brains.***

We want to find out if the quantum eraser will still be operative after our brains are biologically dead. If the quantum eraser doesn't work after that point, then the information placed in our brains while we were alive must be stored somewhere outside the brain.

The Experiment

The proof of an immortal soul would be definitive if we could communicate with a person's accumulated thoughts after experiencing death. However, there is something unsavory about trying to communicate with a deceased person's spirit, so we'll use quantum erasure to test for a soul.

We'll stick to the scientific notion of having a living person's mind collapse the Schrodinger's wave of an object and see if stays collapsed after the person dies. If the reality created by a collapsed Schrodinger's wave goes away after the person dies, then we might assume that the information is elided from the Universe at brain death, and that no soul exists. If the reality remains after the person dies, and subsequent observers see it as the deceased observer saw it, then the information is persisting somewhere, perhaps in an immortal soul. We could verbalize the experiment thus:

"If a person observes an object, and, as a result of observation, collapses its Schrodinger's wave into a set of known properties; and if that knowledge is never recorded on any memory-storing electromechanical device, and is never communicated to anyone else; will the knowledge remain part of the Universe after the person dies?"

The essence of this experiment is to discern the reality that the deceased observer saw, without having ever seeing it ourselves. The Double Slit / Quantum Erasure Experiment allows us to infer a past reality and to surmise whether it is still present in the Universe. I propose the following apparatus:

1. An electron generator.
2. A barrier with a double slit.
3. Two electron detectors, to simplify the ability of our human observers to remember the which-path information. If the electron's Schrodinger's wave collapses through the top slit, then a "0" indicator will light up. If it passes through the bottom slit, then a "1" will light up. We will use these numbers to create a which-path memory for each electron.
4. An "electron trap" cylinder is placed behind the detectors to slow electrons down to a crawl ***after*** they have been detected, taking up to 100 years to traverse the cylinder. This might be a cryogenic device or a device with an electrically charged exterior that controls electron speed by varying the charges that attract and repel electrons, thereby pulling the electrons through the cylinder at a measured speed.
5. A measurement screen to display the electrons as an interference or clump pattern when they emerge from the electron trap.

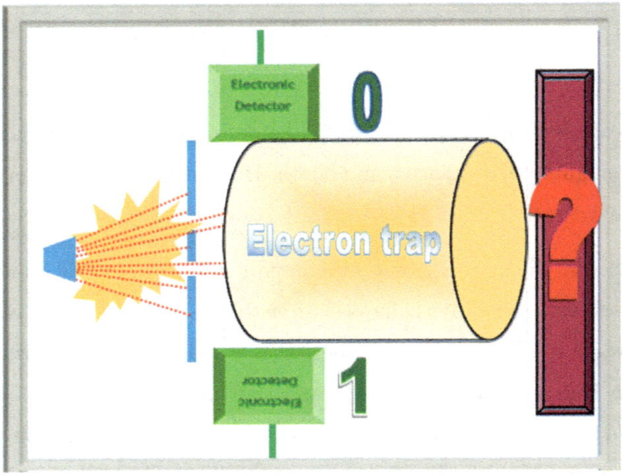

We will manufacture 100 of these devices, numbered 1 to 100. We will set the electron craw speed on each device to the number of years corresponding to its number. Electrons, after bein detected by the top or bottom detector, will transit the "electron trap" in apparatus #1 in one year, # in two years, on up to #100 in 100 years and impact the measurement screen in clumps o interference patterns.

Electrons detected going through the top slit will be assigned a numeric value of "0. Electrons detected going through the bottom slit will be assigned a numeric value of "1." These values will be known immediately, as soon as the electron is fired. A ten-digit binary number will b created in 1's and 0's from the first ten detected. When these ten electrons have passed the detectors the binary number will be converted to decimal and memorized by five observers.

Thus, ten-digit binary numbers will be generated in the range 0000000000 to 1111111111 These correspond to decimal numbers of 0 to 1023. For example, 1001101010 can be remembered a "618." The number generated may be used to deconstruct the "which-path" information of the te electrons that will emerge at the far end of the electron trap cylinder in 1 to 100 years. When al observers have committed the "which-path number" to memory, the digital record will be erased. Th number representing which-path information will reside only in the brains of five people for each o the 100 machines. The test for the soul is whether an interference or clump pattern will emerge o the detection screens of the devices in 1,2,3...100 years, after the electrons have completed their slov walk through the electron trap cylinder.

During these intervening years, our observers will pass away, one by one. The knowledge o the "which-path number" will be erased from their brains as they pass. We will correlate th patterns that emerge on the detection screens with their passing. For device #1, we will expect th electrons to emerge in a clump pattern, as it is likely that most or all observers will still be alive t carry the numeric coded value of the "which-path" information for each electron in their heads. But what results will we get for device #100? Presumably, all the observers will be deceased in 100 years. The "which-path" information in their brains will be destroyed. Will we get an interference pattern by quantum erasure? Or will the clump pattern still be there, indicating that their knowledge has not been lost from the Universe?

We can also check out the intermediate data points, between 1 and 100. If, say, in 40 years, there happens to be one person left in a group when the electrons emerge from device #40, we will expect a clump pattern to emerge. If, say in 45 years, all five observers of device #45 have passed away, and we get an interference pattern, then we might conclude that the which-path knowledge has been lost to the Universe at the time of the last person's passing. If a clump pattern manifests, we might assume that the knowledge still resides somewhere in the Universe, perhaps reposed in the immortal souls of the deceased.

If all devices having at least one surviving observer show "clump" patterns when their electrons emerge from the far end of the electron trap, and all devices having no surviving members show interference patterns when their electrons emerge, then we might assume that the information about the electron paths is not reposed in an immortal soul.

Conclusions

I hope this book has imparted a broad view of the scientific knowledge, unresolved controversies, and metaphysical questions surrounding quantum mechanics and its various interpretations. After writing the first edition, I discovered that the central idea of this book has been independently conceived by *Many Worlds* maven David Deutsch. He conceived an experiment to test for the existence of multiple copies of himself in his theoretical Multiverse.

He discusses the experiment in Chapter 6 of ***The Ghost in the Atom*** that I reference in "Further Reading." The only difference in principle between my proposed experiment and his is that I am proposing test for survival of the soul that transcends death, while he proposes to test for multiple copies of his living self "in other universes."

As to general quantum principles, I reject the *Many Worlds* Theory because I do not believe that a change in the state of one subatomic particle can chain upward to spin off another entire universe, each of whose particles is spinning up more universes. I reject the notion that our decisions, including spontaneous ones such as whether to pull a snack out of the pantry, spins up myriads of universes --- one universe with the snack eaten, the other with it remaining in the pantry --- and if I don't eat it now, separate Universes will be ginned up every time I think about eating it, because a different decision point of whether to eat it or not is reached at a different time in the evolution of the Universe.

Other physicists want us to believe our minds are pre-programmed in a "Super-Deterministic" way by event chains going all the way back to the Big Bang and therefore our decisions are pre-made for us by the Universe such that we have no free will. They believe Super-Determinism would also apply to inanimate actions, such that particles interacting with each other today are doing so because their initial trajectories were defined at the Big Bang. I discount this theory because the Universe has a random aspect, with quantum mechanics at its heart a probabilistic, not deterministic process.

My view is that our consciousness is a phenomenon of one evolving Universe that neither spins off new Universes nor was pre-programmed at the Big Bang to make decisions in the present moment. We cannot know whether it is a purely mechanical effect of molecules and energies in our brains, or whether it has a higher spiritual state of quantum information that survives brain death. Perhaps one day an experiment such as I have suggested will tell us. For now, I lean toward the "Copenhagen" interpretation of QM that consciousness is sufficient but not exclusively necessary to collapse Schrodinger's waves into realities. As Werner Heisenberg states:

*The observer has, rather, only the function of registering decisions, i.e., processes in space and time, and **it does not matter whether the observer is an apparatus or a human being;** but the registration, i.e., the transition from the "possible" to the "actual," is absolutely necessary here and cannot be omitted from the interpretation of quantum theory.*

One of the most popular interpretations gaining traction is known as *Objective (or Spontaneous) Collapse.* It is based on the theory of "decoherence" --- the idea that objects come into existence by their frictional interactions with other particles, just as an ocean wave comes into visible existence as it grates on the sand below it, begins to slow, piling water on top of itself. This leads us to the emerging topic of Quantum Field Theory, postulating that space is permeated by indefinite probability fields for every potential particle, and the particles are materialized from the fields by observation, measurement, interaction, or perhaps simply materialize spontaneously if left alone long enough. In the laboratory, we can isolate particles individually so as to prevent them from spontaneously collapsing, but large objects, containing quintillions of particles should collapse spontaneously, like a nuclear chain reaction, as soon as any one of them collapses to begin it, so that superposition should not be possible for large objects. However, we have not seen any evidence of spontaneous collapse as we increase the numbers of particles in objects we are able to keep in superposition without collapsing

Nevertheless, it seems that quantum mechanics never gets entirely away from the idea that consciousness may be a special factor in creating reality. This is because objects seem to appear spontaneously not only by interaction with other particles, but by the gleaning of information by inference. In the Double Slit Experiment we did not need to detect the electron directly by measuring its electric field to determine which slit it moved through. We can place the detector on one slit and fire an electron. If the electron goes through the slit without the detector and impacts the measurement screen, we infer it passed through that slit because it ***wasn't*** detected. The inference of information seems to be as potent as direct detection in creating the particle. But is information alone sufficient to create the particle, or must there be some other agent that can process the information to make it potent? Can the agent that processes the information be other inanimate objects, or does it require consciousness of some sort to bring it into being?

As this second edition is being written, an experiment has bolstered Eugene Wigner's theory that consciousness is the ultimate arbiter of reality:

https://www.livescience.com/65029-dueling-reality-photons.html

More Than One Reality Exists (in Quantum Physics)

Can two versions of reality exist at the same time? Physicists say they can — at the quantum level, that is.

This perplexing idea was the brainchild of Eugene Wigner, winner of the Nobel Prize for Physics in 1963. In 1961, Wigner had introduced a thought experiment that became known as "Wigner's friend." It begins with a photon — a particle of light. When an observer in an isolated laboratory measures the photon, they find that the particle's polarization — the axis on which it spins — is either vertical or horizontal.

For decades, Wigner's mind-bending proposal was just an interesting thought experiment. But in recent years, important advances in physics finally enabled experts to put Wigner's proposal to the test, Ringbauer said.

"Theoretical advances were needed to formulate the problem in a way that is testable. Then, the experimental side needed developments on the control of quantum systems to implement something like that," he explained.

As our technical ability to run quantum experiments improves, we can verify more aspects of the theory, such as the "Wigner's Friend" experiment described above. Perhaps we'll verify or debunk quantum immortality of human consciousness. We'll then have to be prepared to withstand the societal implications of eliminating the uncertainty.

As for the hype surrounding quantum computing, there is as yet no indication that it will solve any previously unsolvable problems other than arcane ones like computing the factors of prime numbers used for encryption. It may take decades to perfect the technologies required to manufacture quantum computers, so their full capabilities, or lack thereof, are obscure. However, there are so many unknown aspects of quantum mechanics, even after more than a hundred years of intensive theorizing and experimentation, that we cannot predict how quantum technologies will evolve.

Feedback: alsnewideas@gmail.com

Further Study

QM Videos

The advent of software capable of generating videos to explain mathematics and science is one of our generation's most useful technologies. Some excellent QM and Physics videos have been posted to Youtube in the last couple of years. I think the best of the new 2020 and 2021 videos are by Physicist Arvin Ash:

His website:

http://arvinash.com/

Do we even understand quantum mechanics?

https://www.youtube.com/watch?v=OjrEudqgZ1M

Quantum Field Theory

https://www.youtube.com/watch?v=jlEovwE1oHI

Here are some earlier Youtube videos I judge to be within the boundaries of proper scientifi discussion of quantum mechanics but may stretch their points into controversial aspects of quantun metaphysics. You'll have to make your own determinations as to how credible they are.

1 NASA physicist Tom Campbell explains the Double Slit Experiment's intersection o observation, information, and consciousness, especially his view of what happens to realities create by our lifetime observations when we die:

https://www.youtube.com/watch?v=BhMIz_iJtzQ

2 Quantum Mechanics explained by Eugene Khutoryansky.

https://www.youtube.com/watch?v=yahTYvFyv8g

3 This next Eugene Khutoryansky video focuses on quantum erasure, and what happen when information is removed from the brain:

https://www.youtube.com/watch?v=SzAQ36b9dzs

4 A fascinating PBS video on quantum erasure:

https://www.youtube.com/watch?v=8ORLN_KwAgs

5 An insightful conversation between quantum physicists Henry Stapp and John Hagelin:

https://www.youtube.com/watch?v=hSv0NLSCEYo

6 Another Eugene Khutoryansky video. It's not about quantum mechanics per se, but rather explains how light and other frequencies of electromagnetic radiation move through space as particles that gin up electromagnetic forces as they move:

https://www.youtube.com/watch?v=9Tm2c6NJH4Y

Here is also a video of Sean Carroll explaining the *Many Worlds* theory:

https://www.youtube.com/watch?v=zStPYpZk2PY

Metaphysics Consciousness Videos

For a skeptical but open-minded view of metaphysics I recommend the ***Closer to the Truth*** videos that Robert Lawrence Kuhn has done for PBS. He explores the intersection of quantum mechanics with consciousness and theology. He talks to people with a range of viewpoints from the aggressively atheist *Many Worlds* maven Sean Carroll to some physics-trained theologians. I perceived it to be an intellectually unbiased look at the topics of life, science, and religion that confound us as much as they did Blaise Pascal, Charles Darwin, and Erwin Schrodinger:

https://www.youtube.com/watch?v=DmnntDSXrjU

https://www.youtube.com/watch?v=ekfG-PCk25g

https://www.youtube.com/watch?v=5T6VKwaR_vc

https://www.youtube.com/watch?v=hNmfdiSD5qI

https://www.youtube.com/watch?v=sRYdZ25zSjM

https://www.youtube.com/watch?v=0Pt1PT8DsRw

https://www.youtube.com/watch?v=40gymxW9D-c

https://www.youtube.com/watch?v=bI8XiPVmiY4

https://www.youtube.com/watch?v=5xioPAlymFM

https://www.youtube.com/watch?v=UDTwSiBlIhw

https://www.youtube.com/watch?v=rSngHyD44OY

https://www.youtube.com/watch?v=0bEBFdTfHbg

Popular QM Books

I believe these books contain the best explanations of quantum science that a popular audience can enjoy:

Quantum Enigma by Bruce Rosenblum and Fred Kuttner (Oxford University Press, 2011) is a lucid scientific writing, written for laypersons. It covers the spectrum of QM from its philosophy to its physics.

https://www.amazon.com/Quantum-Enigma-Physics-Encounters-Consciousness/dp/0199753814/

The Ghost in the Atom Edited by P. C. W. Davies (Cambridge University Press, 1986) is a classic work going right to the heart of "quantum weirdness." It's a compilation of interviews with some of the great quantum theoretical physicists of the late 20th Century, including John Bell, John Wheeler, and David Bohm. An excellent follow-up to ***Quantum Enigma:***

https://www.amazon.com/Ghost-Atom-Discussion-Mysteries-Quantum/dp/0521457289/

Through Two Doors at Once by Anil Ananthaswamy (Dutton, 2018) explains the Double Slit Experiment from its origin in 1801 to the present. The remarkable aspect is how this experiment remains the core of our understanding of the intersection of quantum mechanics and consciousness. It has been updated with improved equipment that now involves "entangled photons, delayed choice, and quantum erasure."

https://www.amazon.com/Through-Two-Doors-Once-Experiment-ebook/dp/B079KTT95K/

The Quantum Universe by Brian Cox and Jeff Forshaw (Da Capo Press,2013) is ideal for a layperson who desires some scientific heft in popular prose. It explains the quantum mathematics of why particles do not have realities until they are measured, and why measuring one of the properties of a particle, such as its momentum, will cause the particle to de-materialize in one location and re-materialize in another.

https://www.amazon.com/Quantum-Universe-Anything-That-Happen/dp/0306821443/

Why does $E = mc^2$ by Brian Cox and Jeff Forshaw (Da Capo Press, 2010), in my view the best book written on Relativity for a popular audience. Although it touches lightly on quantum mechanics, it presents the framework of relativistic physics that surrounds QM.

https://www.amazon.com/Why-Does-mc2-Should-Care/dp/0306818760/

What is Life? is Erwin Schrodinger's explanation of how quantum mechanics intersects with life. It was written in 1942, more than a decade before the discovery of DNA. Even though dated, it provides wonderful insights into the physics of life and the controversies of quantum mechanics related to life and consciousness. It's also a revealing insight into how such an eminent scientist as Schrodinger views many topics of general interest.

https://www.amazon.com/What-Life-Autobiographical-Sketches-Classics/dp/1107604664/

Science Set Free by Rupert Sheldrake (Deepak Chopra September 3, 2013) is considered speculative pseudo-science. However, Sheldrake has worked closely with esteemed quantum physicist David Bohm who shares some of his views. Sheldrake postulates a higher quantum knowledge residing outside the bodies of organisms that drives their evolution and stores the collective knowledge acquired by the species as it evolves.

https://www.amazon.com/Science-Set-Free-Paths-Discovery/dp/0770436722/

Who's in Charge by Michael S. Gazzinga (Ecco; Reprint edition, 2011) explores the current state of brain science. It touches lightly on how quantum information may impact the processing of consciousness in the brain, while explaining how most scientists view the brain and its generation of consciousness.

https://www.amazon.com/Whos-Charge-Free-Science-Brain-ebook-dp-B005UD1EVG/dp/B005UD1EVG/

The Information Philosopher

This website discusses the information theory of quantum mechanics. This is the theory that information created by observation causes Schrodinger's waves to collapse into reality. In this interpretation, consciousness plays no special role in creating information that an inanimate measuring apparatus would not also create, but consciousness is as much a store of information as a device, and therefore is sufficient to run our soul-searching experiment:

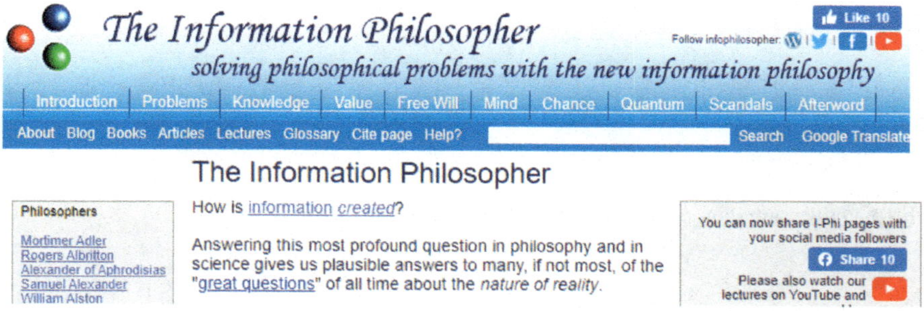

https://www.informationphilosopher.com/

https://www.informationphilosopher.com/quantum/observer/

About the Author

https://www.amazon.com/stores/Alan-Sewell/author/B00557PQDY

"Understanding History is a key to understanding the present and extrapolating the future" Alan Sewell

Although every day is a new day, the new days are layered on top of repeating cycles o history as old as Mankind. I've analyzed historical and current events and applied their historica lessons to today's business and economic issues.

I've been commended for "interpreting the American experience" into six critical watershe periods when political and economic crises were resolved in order to advance the country into th next era of its history:

FRAGMENTATION - FEDERALISM - UNION (1783 - 1815)

SECESSION - WAR - NATIONALISM (1861 - 1868)

WEALTH - DEPRESSION - EXPANSION (1890-1900)

WEALTH - DEPRESSION - LIBERALISM (1925-1933)

CHAOS - HUMILIATION - CONSERVATISM (1968-1981)

INSTABILITY --- RECESSION ---- {Populism or Progressivism} (2008-?)

I've also been commended on writing reviews of books on science and the history of science. began with the study of science and engineering in university, while writing articles for popula national history magazines, then pursued a career developing algorithms to manage the business c multinational corporations. I have returned to my specialization of presenting history, science, an economics to popular audiences as easily comprehended books and articles, without omitting the cor tenets and controversies of the subjects. I am a popular commenter in the Wall Street Journal on line edition on topics of business, economics, history, and science.

My newest scientific book is ***Einstein's Car***, a unique perspective on Relativity:

https://www.amazon.com/Einsteins-Car-Relativity-Alan-Sewell-ebook/dp/B0CBKY7HW1

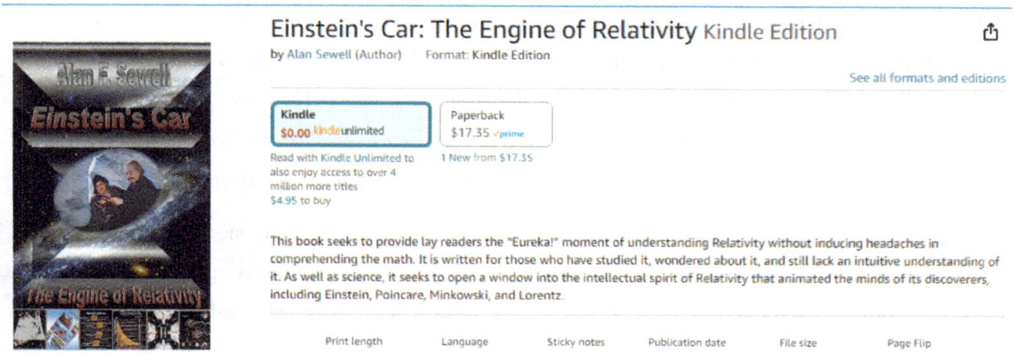

Albert Einstein and his second wife Elsa

This book seeks to provide lay readers the "Eureka!" moment of understanding Relativity without inducing headaches in comprehending the math. It is written for those who have studied it, wondered about it, and still lack an intuitive understanding of it. As well as science, it seeks to open a window into the intellectual spirit of Relativity that animated the minds of its discoverers, including Einstein, Poincare, Minkowski, and Lorentz.

Reading about Relativity as a non-professional physicist is frustrating. You feel it's an important aspect of the Universe you want to understand. But studying it either balks you with incomprehensible mathematical formulas or tries to placate you with platitudes of "everything is relative except the invariant speed of light" without explaining the how's and why's.

No matter how much information you take in, understanding remains elusive. You find yourself immersed in allegories such as Einstein's Train and Embankment, The Twins Paradox, and even The Andromeda Paradox, explained in contradictory ways. How can we glean an intuitive understanding of how time behaves in Relativity, when in everyday life we only experience time ticking at a constant rate and events pegged to a universally agreed upon timeline?

My study of Relativity began with the popular cosmology and physics books of the 1970s by authors like George Gamow and Isaac Asimov. My high school science project was the visual representation of Relativity's four dimensional Spacetime. With the advent of computers, I programmed three dimensional coordinates in engineering applications and four-dimensional Lorentzian coordinate systems in the study of Relativity. After a career in business centered around developing algorithms to control operations of multinational companies, I began writing on topics of history and economics.

I returned to the study of Relativity for four and a half years, wanting to visualize it as a holistic entity, rather than a mathematical abstraction. It would be pointless, and even insulting, to readers to merely rehash material available in popular physics books, the Internet, and Youtube videos, which after a lifetime of study I found remarkably unenlightening.

I restudied the theory from the original writings of Einstein and its other founders to the current state of information and belief. I progressed from the superficial familiarity gleaned from reading popular physics books to an intuitive understanding, the way one progresses in learning a new language --- from tedious thinking of how to translate it word by word, to intuitively hearing and speaking it without conscious translation.

My discussions with Relativity physicists lead me to believe my insights are worthwhile. I've been educated to aspects of Relativity I didn't understand, while helping physicists understand some so-called paradoxes by looking at Relativity in a broader way than its traditional dogma is taught.

Made in United States
North Haven, CT
10 November 2026

22096031R00046